Selected Titles in This Series

D1684566

(Continued in the back of this publication)

Homogeneous Spaces, Tits Buildings, and Isoparametric Hypersurfaces

MEMOIRS
of the
American Mathematical Society

Number 752

Homogeneous Spaces, Tits Buildings, and Isoparametric Hypersurfaces

Linus Kramer

July 2002 • Volume 158 • Number 752 (third of 4 numbers) • ISSN 0065-9266

American Mathematical Society
Providence, Rhode Island

2000 *Mathematics Subject Classification.*
Primary 51Hxx, 53Cxx; Secondary 51E12, 57T15.

Library of Congress Cataloging-in-Publication Data

Kramer, Linus, 1964–
Homogeneous spaces, Tits buildings, and isoparametric hypersurfaces / Linus Kramer.
p. cm. — (Memoirs of the American Mathematical Society, ISSN 0065-9266 ; no. 752)
"Volume 158, number 752 (third of 4 numbers)."
Includes bibliographical references.
ISBN 0-8218-2906-8 (alk. paper)
1. Buildings (Group theory) Global differential geometry. 3. Finite generalized quadrangles. 4. Homogeneous spaces. I. Title. II. Series.

QA3 .A57 no. 752
[QA174.2]
510 s—dc21
[512′.2] 2002018395

Memoirs of the American Mathematical Society

This journal is devoted entirely to research in pure and applied mathematics.

Subscription information. The 2002 subscription begins with volume 155 and consists of six mailings, each containing one or more numbers. Subscription prices for 2002 are $524 list, $419 institutional member. A late charge of 10% of the subscription price will be imposed on orders received from nonmembers after January 1 of the subscription year. Subscribers outside the United States and India must pay a postage surcharge of $31; subscribers in India must pay a postage surcharge of $43. Expedited delivery to destinations in North America $35; elsewhere $130. Each number may be ordered separately; *please specify number* when ordering an individual number. For prices and titles of recently released numbers, see the New Publications sections of the *Notices of the American Mathematical Society.*

Back number information. For back issues see the *AMS Catalog of Publications.*

Subscriptions and orders should be addressed to the American Mathematical Society, P. O. Box 845904, Boston, MA 02284-5904. *All orders must be accompanied by payment.* Other correspondence should be addressed to Box 6248, Providence, RI 02940-6248.

Memoirs of the American Mathematical Society is published bimonthly (each volume consisting usually of more than one number) by the American Mathematical Society at 201 Charles Street, Providence, RI 02904-2294. Periodicals postage paid at Providence, RI. Postmaster: Send address changes to Memoirs, American Mathematical Society, P. O. Box 6248, Providence, RI 02940-6248.

This publication is indexed in *Science Citation Index®*, *SciSearch®*, *Research Alert®*, *CompuMath Citation Index®*, *Current Contents®/Physical, Chemical & Earth Sciences.*
Printed in the United States of America.

∞ The paper used in this book is acid-free and falls within the guidelines established to ensure permanence and durability.
Visit the AMS home page at URL: `http://www.ams.org/`

10 9 8 7 6 5 4 3 2 1 07 06 05 04 03 02

Contents

Abstract

We classify 1-connected compact homogeneous spaces which have the same rational cohomology as a product of spheres $\mathbb{S}^{n_1} \times \mathbb{S}^{n_2}$, with $3 \leq n_1 \leq n_2$ and n_2 odd. As an application, we classify compact generalized quadrangles (buildings of type C_2) which admit a point transitive automorphism group, and isoparametric hypersurfaces which admit a transitive isometry group on one focal manifold.

Received by the editor February 22, 2001.

1991 *Mathematics Subject Classification.* Primary 51H, 53C; Secondary 51E12, 57T15.

Key words and phrases. Buildings, generalized quadrangles, topological geometry, isoparametric submanifolds, polar representations, homogeneous spaces, Stiefel manifolds.

Supported by a Heisenberg Fellowship by the Deutsche Forschungsgemeinschaft.

Introduction

The classification of compact Lie groups acting transitively on spheres due to Montgomery, Samelson and Borel was one of the main achievements in the early theory of compact transitive Lie transformation groups. Later, Hsiang and Su classified compact transitive groups on (sufficiently highly connected) Stiefel manifolds. Their results were extended by Scheerer, Schneider and other authors. Given a compact 1-connected manifold X, it is in general difficult to classify all compact Lie groups which act transitively on X. The problem becomes more complicated if only certain homotopy invariants of X such as the cohomology ring are known. If the Euler characteristic of X is positive, then results of Borel, De Siebenthal and Wang can be used. However, if the Euler characteristic is 0, there is no general classification method.

* * *

In this book we classify all 1-connected homogeneous spaces G/H of compact Lie groups which have the same rational cohomology as a product of spheres

$$\mathbb{S}^{n_1} \times \mathbb{S}^{n_2},$$

with $3 \leq n_1 \leq n_2$ and n_2 odd. Note that this implies that the Euler characteristic of G/H is 0. Examples of such spaces are — besides products of spheres — Stiefel manifolds of orthonormal 2-frames in real, complex, or quaternionic vector spaces; another class of examples are certain homogeneous sphere bundles. The following theorem is a direct consequence of this rational classification.

Theorem *Let $X = G/H$ be a 1-connected compact homogeneous space of a compact connected Lie group G. Assume that G acts effectively and contains no normal transitive subgroup, and that X has the same integral cohomology as a product of spheres*

$$\mathbb{S}^{n_1} \times \mathbb{S}^{n_2}$$

with $3 \leq n_1 \leq n_2$ and n_2 odd. There are the following possibilities for G/H and the numbers (n_1, n_2).
(1) Stiefel manifolds

$$\begin{array}{ll} \mathrm{SO}(2n)/\mathrm{SO}(2n-2) = V_2(\mathbb{R}^{2n}) & (2n-2, 2n-1) \\ \mathrm{SU}(n)/\mathrm{SU}(n-2) = V_2(\mathbb{C}^n) & (2n-3, 2n-1) \\ \mathrm{Sp}(n)/\mathrm{Sp}(n-2) = V_2(\mathbb{H}^n) & (4n-5, 4n-1). \end{array}$$

(2) Certain homogeneous sphere bundles

$$\begin{array}{ll} \mathrm{Sp}(n) \times \mathrm{Sp}(2)/\mathrm{Sp}(n-1) \cdot \mathrm{Sp}(1) & (7, 4n-1) \\ \mathrm{Sp}(n) \times \mathrm{SU}(3)/\mathrm{Sp}(n-1) \cdot \mathrm{Sp}(1) & (5, 4n-1) \\ \mathrm{Sp}(n) \times \mathrm{Sp}(2)/\mathrm{Sp}(n-1) \cdot \mathrm{Sp}(1) \cdot \mathrm{Sp}(1) & (4, 4n-1). \end{array}$$

(3) Products of homogeneous spheres

$$K_1/H_1 \times K_2/H_2 = \mathbb{S}^{n_1} \times \mathbb{S}^{n_2}$$

where K_1/H_1 is one of the spaces

$$\mathrm{SO}(n)/\mathrm{SO}(n-1) = \mathbb{S}^{n-1}$$
$$\mathrm{SU}(n)/\mathrm{SU}(n-1) = \mathbb{S}^{2n-1}$$
$$\mathrm{Sp}(n)/\mathrm{Sp}(n-1) = \mathbb{S}^{4n-1}$$
$$\mathrm{G}_2/\mathrm{SU}(3) = \mathbb{S}^{6}$$
$$\mathrm{Spin}(7)/\mathrm{G}_2 = \mathbb{S}^{7}$$
$$\mathrm{Spin}(9)/\mathrm{Spin}(7) = \mathbb{S}^{15}$$

and K_2/H_2 is one of the spaces

$$\mathrm{SO}(2n)/\mathrm{SO}(2n-1) = \mathbb{S}^{2n-1}$$
$$\mathrm{SU}(n)/\mathrm{SU}(n-1) = \mathbb{S}^{2n-1}$$
$$\mathrm{Sp}(n)/\mathrm{Sp}(n-1) = \mathbb{S}^{4n-1}$$
$$\mathrm{Spin}(7)/\mathrm{G}_2 = \mathbb{S}^{7}$$
$$\mathrm{Spin}(9)/\mathrm{Spin}(7) = \mathbb{S}^{15}$$

(4) Some sporadic spaces

$$\begin{array}{cl}
\mathrm{E}_6/\mathrm{F}_4 & (9,17) \\
\mathrm{Spin}(10)/\mathrm{Spin}(7) & (9,15) \\
\mathrm{Spin}(9)/\mathrm{G}_2 = V_2(\mathbb{O}^2) & (7,15) \\
\mathrm{Spin}(8)/\mathrm{G}_2 = \mathbb{S}^7 \times \mathbb{S}^7 & (7,7) \\
\mathrm{SU}(6)/\mathrm{Sp}(3) = \mathrm{SU}(5)/\mathrm{Sp}(2) & (5,9) \\
\mathrm{Spin}(10)/\mathrm{SU}(5) = \mathrm{Spin}(9)/\mathrm{SU}(4) & (6,15) \\
\mathrm{Spin}(7)/\mathrm{SU}(3) = V_2(\mathbb{R}^8) & (6,7) \\
\mathrm{Sp}(3)/\mathrm{Sp}(1) \times \mathrm{Sp}(1) & (4,11) \\
\mathrm{Sp}(3)/\mathrm{Sp}(1) \times {}^{\mathbb{H}}\rho_{3\lambda_1}(\mathrm{Sp}(1)) & (4,11) \\
\mathrm{SU}(5)/\mathrm{SU}(3) \times \mathrm{SU}(2) & (4,9).
\end{array}$$

The proof proceeds as follows. We show first that G/H has the same rational homotopy groups as the product $\mathbb{S}^{n_1} \times \mathbb{S}^{n_2}$. This homotopy theoretic result follows from a generalization of a theorem by Cartan and Serre. The rational homotopy groups of a compact Lie group G can be determined explicitly; they depend only on the Dynkin diagram of G and the rank of the central torus. In particular, a compact connected Lie group has the same rational homotopy groups as a product of odd-dimensional spheres. It follows in our situation that $\mathrm{rk}(G) - \mathrm{rk}(H) \in \{1, 2\}$, depending on whether n_1 is even or odd. Using this fact, we determine the rational Leray-Serre spectral sequence of the principal bundle

$$H \longrightarrow G \longrightarrow G/H.$$

If n_1 is odd, then the spectral sequence collapses, and G has the same rational cohomology as $H \times \mathbb{S}^{n_1} \times \mathbb{S}^{n_2}$. If n_1 is even, then there are non-zero differentials in the spectral sequence and the situation is more complicated. In both cases we obtain a relation between the numbers (n_1, n_2) and the degrees of the primitive elements in the rational cohomology of G and H.

It follows also that G has a semisimple normal transitive subgroup $K \trianglelefteq G$ which has at most two almost simple factors. By a general result about transitive permutation groups it suffices to determine the pair $(K, K \cap H) \subseteq (G, H)$ in order to determine all possibilities for the larger group G. Replacing G by K we may therefore assume that G is almost simple or that G is semisimple with two almost simple factors. The condition $n_1 \geq 3$ guarantees then that H is also semisimple. We determine all such pairs (G, H) with the right rational cohomology and all possible embeddings $H \hookrightarrow G$ using representation theory. Thus, we obtain an explicit classification of these homogeneous spaces together with the transitive group actions.

In the special case that G/H has the same integral cohomology as $\mathbb{S}^{n_1} \times \mathbb{S}^{n_2}$ we obtain the list of homogeneous spaces given in the theorem above. In the course of the proof we determine also all compact connected Lie groups which act transitively on 1-connected rational homology spheres; in particular, we reprove the classification of transitive actions on spheres and on spaces which have the same homology as the Stiefel manifolds $V_2(\mathbb{R}^{2n+1})$.

* * *

We apply our result to a problem in submanifold geometry. A closed hypersurface in a sphere is called isoparametric if its principal curvatures are constant. Hsiang and Lawson classified all isoparametric hypersurfaces which admit a transitive group of isometries; these homogeneous isoparametric hypersurfaces arise as principal orbits of isotropy representations of non-compact symmetric spaces of rank 2.

By a result of Münzner, the number of distinct principal curvatures of an isoparametric hypersurface is $g = 1, 2, 3, 4, 6$; the hypersurfaces with $g = 1, 2, 3$ have been classified in the 30s by Segre and Cartan. Some hypersurfaces with $g = 6$ were classified by Dorfmeister and Neher; the full classification for $g = 6$ still seems to be an open problem. The case of hypersurfaces with $g = 4$ is much more difficult. Takagi proved uniqueness for isoparametric hypersurfaces with $g = 4$ distinct principal curvatures and multiplicities $(1, k)$. On the other hand, Ferus, Karcher and Münzner showed that there are many non-homogeneous isoparametric hypersurfaces with $g = 4$ distinct principal curvatures, and Stolz recently obtained sharp number theoretic restrictions on the possible dimensions of such hypersurfaces. Some of the known inhomogeneous examples have homogeneous focal manifolds, so the question arises if the classification by Hsiang and Lawson can be generalized to transitive actions on focal manifolds. In view of the classification by Segre, Cartan, Dorfmeister and Neher, the cases $g = 1, 2, 3$ are less interesting. If $g = 4$ and if the multiplicities (m_1, m_2) of the isoparametric hypersurface are large enough, then the focal manifolds have the same integral cohomology as a product of spheres. We apply our result about homogeneous spaces to this problem and obtain a complete classification.

Theorem *Let M be an isoparametric hypersurface with 4 distinct principal curvatures. Suppose that the isometry group of M acts transitively on one of the focal manifolds, and that this focal manifold is 2-connected. Then either the hypersurface itself is homogeneous (and explicitly known), or it is of Clifford type with multiplicities $(8, 7)$ or $(3, 4k - 4)$.*

Recently, Wolfrom showed in his Ph.D. Thesis that the theorem above holds also if one drops the assumption that the focal manifold is 2-connected, and proved a similar result for $g = 6$. The final result is as follows.

Theorem *Let M be an isoparametric hypersurface, and suppose that the isometry group of M acts transitively on one of the focal manifolds. Then either the hypersurface itself is homogeneous (and explicitly known), or it is of Clifford type with multiplicities* $(8, 7)$ *or* $(3, 4k - 4)$.

This theorem gives in particular a new, independent proof for the classification by Hsiang and Lawson.

* * *

Another application is in topological geometry. Every isotropic simple algebraic group, in particular every non-compact real simple Lie group, gives rise to a spherical Tits building. These buildings are characterized by the so-called Moufang condition. In the case of a non-compact simple Lie group, the building inherits a compact topology from the group action. These compact buildings are closely related to symmetric spaces, Fürstenberg boundaries and isoparametric submanifolds. They play also a rôle in the theory of Hadamard spaces and rigidity results.

Tits classified all irreducible spherical buildings of rank at least 3 by showing that they automatically satisfy the Moufang property. In contrast to this generalized polygons, i.e. spherical buildings of rank 2, need not be Moufang, and there is no way to classify them without further assumptions.

In view of the examples above, it is natural to consider compact generalized polygons and to try to classify them in terms of their automorphism groups. A result of Knarr and the author states that such a building is of type A_2, C_2, or G_2. Topologically, a compact generalized polygon looks very similar to an isoparametric foliation with $g = 3, 4, 6$ distinct principal curvatures, respectively; in particular, the cohomology of these spaces can be determined. This is the analogue of Münzner's theorem mentioned above.

An A_2-building is the same as a projective plane; all compact homogeneous projective planes have been classified by Löwen and Salzmann. The compact homogeneous G_2-buildings have been classified by the author. The remaining cases are the generalized quadrangles, i.e. the buildings of type C_2. As in the case of isoparametric hypersurfaces, this is much more involved. The easiest case here are the quadrangles with Euler characteristic 4 which have been classified by the author. The remaining case, namely compact quadrangles of Euler characteristic 0, is very interesting, since the inhomogeneous isoparametric hypersurfaces discovered by Ferus, Karcher and Münzner are examples of such (non-Moufang) quadrangles.

By general arguments, a transitive automorphism group on a 1-connected compact quadrangle contains a compact transitive Lie subgroup (the automorphism group of a compact building is in general not compact). Thus, we can apply our result to classify transitive actions of compact Lie groups on compact quadrangles. We obtain a list of all possible transitive actions. For three infinite series we show that the group action determines the quadrangle up to isomorphism. In fact, we can (to some extent) use the same geometric methods and arguments both for isoparametric hypersurfaces with $g = 4$ distinct principal curvatures and for compact quadrangles.

Nevertheless, the classification of the compact quadrangles is much more difficult than the classification of isoparametric hypersurfaces. This is due to the fact that an isoparametric hypersurface sits inside some Euclidean space on which the group acts linearly. Even though this surrounding space also exists for generalized quadrangles, it has no natural Euclidean structure, and it is not a priori clear that the group action has to be linear. Therefore representation theory has to be replaced by arguments about compact transformation groups acting on locally compact spaces. The result is as follows.

Theorem *Let $\mathfrak{G}$ be a compact connected quadrangle. Assume that the point space is 9-connected, and that the automorphism group is point transitive. Then $\mathfrak{G}$ is a Moufang quadrangle (in fact the dual of a classical quadrangle associated to a hermitian form over $\mathbb{R}$, $\mathbb{C}$ or $\mathbb{H}$).*

At this point, we should mention the following related results by Grundhöfer, Knarr and the author.

Theorem *Let Δ be a compact connected irreducible spherical building of rank at least 2. Assume that the automorphism group is chamber transitive. Then Δ is the Moufang building associated to a simple non-compact Lie group G. If $H \subseteq \mathrm{Aut}(\Delta)$ is a connected chamber transitive subgroup, then either $H = G$, or H is compact.*

All compact connected chamber transitive groups in the theorem above were determined by Eschenburg and Heintze. Combining these results, we have the following theorem.

Theorem *Let Δ be a compact connected irreducible spherical building of rank $k \geq 2$. Assume that the automorphism group is transitive on one type of vertices of Δ. If the building is of type C_2 assume in addition that either (1) the vertex space in question is 9-connected, or (2) that the two vertex sets have the same dimension, or (3) that the action is chamber transitive. Then Δ is a Moufang building associated to a simple non-compact real Lie group of real rank k.*

In the C_2-case, the assumption on the connectivity cannot be dropped completely, since there are counterexamples which are not highly connected. In view of the results in the present book, the following conjecture for the C_2-case is very natural.

Conjecture *Let $\mathfrak{G}$ be a compact connected generalized quadrangle. Assume that the automorphism group is point or line transitive. The either $\mathfrak{G}$ is a Moufang quadrangle, or $\mathfrak{G}$ is of Clifford type with topological parameters $(3, 4k)$ or $(7, 8)$.*

Finally, I should mention here the following new and beautiful result by Immervoll: every isoparametric hypersurface with $g = 4$ distinct principal curvatures is a C_2-building.

* * *

The material is organized as follows. In the first chapter we collect some well-known facts about the algebraic topology of fibrations, spectral sequences, and Eilenberg-MacLane spaces. This chapter should be accessible for any reader with basic knowledge about algebraic topology. In Chapter 2 we prove an extension of the Cartan-Serre theorem about rational homotopy groups by a standard homotopy theoretic method. This is applied in Chapter 3 in order to determine the rational Leray-Serre spectral sequence associated to the transitive group action.

All facts about representations of compact Lie groups on real, complex or quaternionic vector spaces which are used in the classification are presented in Chapter 4, which also contains tables about compact almost simple Lie groups and their low-dimensional representations. These facts may be of some independent interest.

Our classification of homogeneous spaces is carried out in Chapters 5 and 6. The classification is stated at the end of Chapter 3. Readers who are willing to accept this result without entering into the details can skip Chapters 5 and 6. However, the representation theory of Chapter 4 is used again in the last two chapters.

In Chapter 7 we classify transitive actions of compact Lie groups on compact quadrangles. The homogeneous focal manifolds of isoparametric hypersurfaces with $g = 4$ distinct principal curvatures are classified in Chapter 8, together with all possible transitive group actions. Here, the reader is assumed to be familiar either with topological geometry or with isoparametric hypersurfaces.

The logical dependencies of the chapters are as follows:

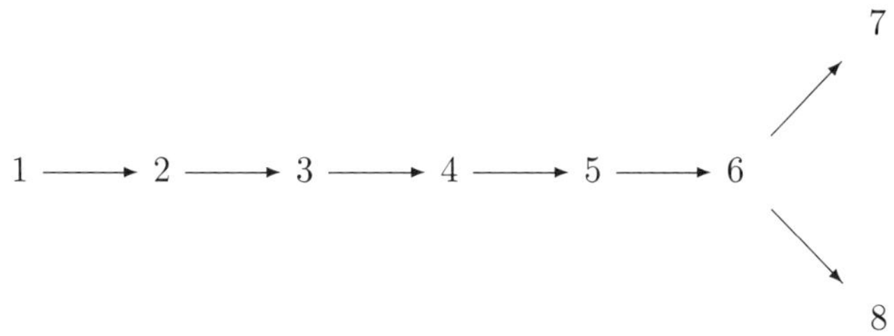

In particular, Chapter 7 (topological geometry) and Chapter 8 (submanifold geometry) are in principle independent of each other. Nevertheless, the subjects of these two chapters, isoparametric hypersurfaces and compact polygons, share many geometric properties which I tried to emphasize. Often, the proofs in Chapters 7 and 8 are quite similar and geometric. Besides the global properties of isoparametric hypersurfaces, very little differential geometry is needed in the classification. I hope that the present book will be useful both for differential geometers and for topological geometers, and that it helps to broaden the bridge between the two fields.

* * *

The book is a revised, corrected and expanded version of my *Habilitationsschrift*. I would like to thank my teachers Theo Grundhöfer and Reiner Salzmann for their constant interest and support. Harald Biller, Oliver Bletz, Norbert Knarr, Gerhard Röhrle, Stephan Stolz, Markus Stroppel, Hendrik Van Maldeghem, and Martin Wolfrom made helpful suggestions or spotted errors. I used a computer program by Richard Bödi and Michael Joswig to check the tables for the representations of simple Lie groups. Robert Bryant and Friedrich Knop helped me with a question about a certain representation. The commutative diagrams in the original manuscript were drawn with Paul Taylors diagrams TeX-package. Finally, I would like to thank my wife, Katrin Tent, not only for reading the manuscript. Without her support, this book would not have been possible.

Gerbrunn, February 2001
Linus Kramer

Man vergilt seinem Lehrer schlecht,
wenn man immer nur der Schüler bleibt.
F. N.

CHAPTER 1

The Leray-Serre spectral sequence

Our main tool from algebraic topology is the Leray-Serre spectral sequence. It relates the cohomology rings of the fibre and the base of a fibration with the cohomology of the total space. Although spectral sequences are standard devices in topology, they tend to be somewhat intimidating to non-specialists. The aim of this chapter is to give a basic introduction to the relevant notions and techniques.

Let us consider a specific example, the Leray-Serre spectral sequence with field coefficients. Let K be a field, and let F, B be topological spaces. The Künneth Theorem asserts that the K-cohomology of the product $E = F \times B$ is given by

$$\mathbf{H}^k(E;K) \cong \bigoplus_{i+j=k} \mathbf{H}^i(F;K) \otimes \mathbf{H}^j(B;K).$$

It is convenient to visualize the K-modules $\mathbf{E}^{i,j} = \mathbf{H}^i(F;K) \otimes \mathbf{H}^j(B;K)$ as distributed on the lattice $\mathbb{Z}^2 \subseteq \mathbb{R}^2$: attached to the point (i,j) is the vector space $\mathbf{E}^{ij}$. Then $\mathbf{H}^k(E;K)$ is obtained by adding up all vector spaces along the line $i + j = k$.

Now suppose that E is not a product, but the total space of a (K-simple) fibre bundle $F \longrightarrow E \longrightarrow B$. Then the cohomology of E is 'smaller' than the cohomology of the trivial bundle $F \times B$. The recipe to obtain the K-modules $\mathbf{H}^k(E;K)$ is as follows. Start with the collection of K-modules $\mathbf{E}_2^{i,j} = \mathbf{H}^i(F;K) \otimes \mathbf{H}^j(B;K)$ as before. There exists a collection of maps, the differentials, denoted $d_2 : E_2^{i,j} \longrightarrow E_2^{i+2,j-1}$. These differentials should be visualized as arrows going from (i,j) to $(i+2, j-1)$.

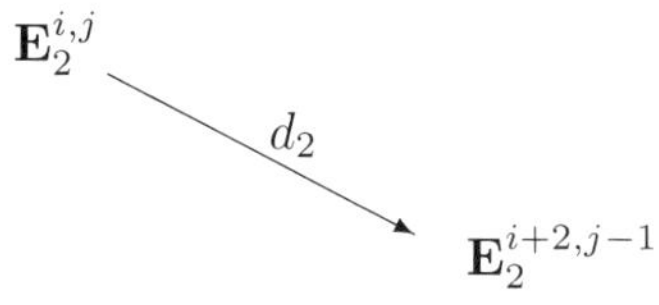

They satisfy the relation $d_2 \circ d_2 = 0$, so we can take their cohomology (kernel mod image) at each point (i,j). Call the resulting K-module $\mathbf{E}_3^{i,j}$. Note that this is a quotient of a submodule of $\mathbf{E}_2^{i,j}$, so its dimension is smaller. Again, these K-modules should be viewed as distributed in the plane. There is another differential d_3, this time from (i,j) to $(i+3, j-2)$. Now this process is iterated *ad infinitum*. The arrows $d_2, d_3, d_4, \ldots$ become longer, and their slope approaches -1. In the limit, one obtains a collection of K-modules denoted $\mathbf{E}_\infty^{i,j}$. Similarly as in the Künneth Theorem one has

$$\mathbf{H}^k(E;K) \cong \bigoplus_{i+j=k} \mathbf{E}_\infty^{i,j}.$$

In fact, one has not to go to infinity in this situation. Since $\mathbf{E}_2^{i,j} = 0$ for $i < 0$ or $j < 0$, the arrows d_r starting or ending at (i,j) are trivial maps for r large enough (e.g. $r > \max\{i,j\}$). Thus, the modules $\mathbf{E}_r^{i,j}$ become stationary after some time.

However, there is one big problem: in general, no information is given about the arrows $d_2, d_3, \ldots$! Thus, it seems to be impossible to determine $\mathbf{E}_3^{i,j}, \mathbf{E}_4^{i,j}, \ldots$. Here, the multiplicative structure of the cohomology becomes important. The arrows d_r act as derivations, and this makes it (often) feasible to determine all terms in the spectral sequence.

The material of this chapter can be found in McCleary [**69**], Borel [**9, 12**], Spanier [**90**], Whitehead [**113**], and Fomenko-Fuchs-Gutenmacher [**35**].

Throughout this chapter, R is a principal ideal domain.

1.A. Additive structure

1.1. Graded and bigraded modules

A *graded R-module* is a direct sum

$$M = M^\bullet = \bigoplus_{i \in \mathbb{Z}} M^i$$

of R-modules M^i, indexed by the integers. Similarly, a *bigraded R-module* is a direct sum

$$M = M^{\bullet\bullet} = \bigoplus_{i,j \in \mathbb{Z}} M^{i,j}$$

of R-modules $M^{i,j}$, indexed by pairs of integers. The elements of M^i or $M^{i,j}$ are called *homogeneous* of *degree* i or *bidegree* (i,j), respectively. A graded or bigraded module is of *finite type* if the M^i or $M^{i,j}$ are finitely generated. A submodule $N \subseteq M^\bullet$ is graded if $N = \bigoplus_{i \in \mathbb{Z}} N^i$, where $N^i = N \cap M^i$.

1.2. Total gradings and tensor products

Associated to a bigraded module $M^{\bullet\bullet}$ is the graded module $M^\bullet = \operatorname{Tot}(M)$ which is graded by the total degree,

$$\operatorname{Tot}(M)^i = \bigoplus_{j+k=i} M^{j,k}.$$

A typical example for a bigraded module is a tensor product of graded modules. Put

$$(M \otimes N)^{i,j} = M^i \otimes N^j.$$

Then the corresponding graded module is

$$(M \otimes N)^i = \bigoplus_{j+k=i} (M^j \otimes N^k).$$

Another example is obtained from filtrations.

1.3. Filtrations and associated gradings

A *filtration* of a module M is a collection of submodules

$$\cdots \subseteq \mathbf{F}^i M \subseteq \mathbf{F}^{i-1} M \subseteq \mathbf{F}^{i-2} M \subseteq \mathbf{F}^{i-3} M \subseteq \cdots$$

The filtration is *bounded* if $\mathbf{F}^0 M = M$, and *convergent* if $\bigcap_{i \geq 0} \mathbf{F}^i M = 0$. If $M = M^\bullet$ is graded, then one requires that the submodules in the filtration are graded. An example of a filtration is the following. Let X be a CW complex, and let $X^{(k)}$ denote its k-skeleton. Then

$$\mathbf{F}^i\mathbf{H}^j(X;R) = \ker[\mathbf{H}^j(X^{(i-1)};R) \longleftarrow \mathbf{H}^j(X;R)]$$

defines a filtration of the cohomology module of X.

Associated to a (bounded and convergent) filtration is the bigraded module $\mathbf{G}(M)$ which is defined by

$$\mathbf{G}(M)^{i,j} = \frac{\mathbf{F}^i M^{i+j}}{\mathbf{F}^{i+1} M^{i+j}}$$

In general, it can be difficult to recover the graded module M from $\mathbf{G}(M)$ (this is a problem about module extensions). However, if $M^\bullet$ is of finite type, and if R is a field, then

$$\mathrm{Tot}(\mathbf{G}(M))^i = \bigoplus_{j\in\mathbb{Z}} \frac{\mathbf{F}^j M^i}{\mathbf{F}^{j+1} M^i} \cong M^i.$$

1.4. Differential graded modules

A map $f : M^\bullet \longrightarrow N^\bullet$ of degree r between graded modules is an R-linear map which increases degrees by r, i.e. $f(M^i) \subseteq N^{i+r}$. A differential is a map $d : M^\bullet \longrightarrow M^\bullet$ of degree 1, with $d^2 = 0$. The *cohomology* of (M, d) is the graded module

$$\mathbf{H}^i(M) = \frac{\ker[M^i \xrightarrow{d} M^{i+1}]}{\mathrm{im}[M^{i-1} \xrightarrow{d} M^i]}.$$

An R-linear map $f : M^{\bullet\bullet} \longrightarrow N^{\bullet\bullet}$ has bidegree (r, s) if $f(M^{i,j}) \subseteq N^{i+r,j+s}$. A *differential* d of bidegree $(r, 1-r)$ is a map $d : M \longrightarrow M$ of bidegree $(r, 1-r)$ with $d^2 = 0$. The *cohomology* of (M, d) is the bigraded module

$$\mathbf{H}^{i,j}(M) = \frac{\ker[M^{i,j} \xrightarrow{d} M^{i+r,j+1-r}]}{\mathrm{im}[M^{i-r,j-1+r} \xrightarrow{d} M^{i,j}]}.$$

Note that d is a differential on $\mathrm{Tot}(M)$, and that $\mathrm{Tot}(\mathbf{H}^{\bullet\bullet}(M)) = \mathbf{H}^\bullet(\mathrm{Tot}(M))$. The pair (M, d) is called a *differential (bi)graded module.*

1.5. Spectral sequences

An $\mathbf{E}_2$-*spectral sequence* is a collection of differential bigraded modules $\mathbf{E}_r$, indexed by $r = 2, 3, 4, \cdots$, endowed with differentials d_r of bidegree $(r, 1-r)$, such that

$$\mathbf{E}_{r+1} \cong \mathbf{H}(\mathbf{E}_r).$$

The spectral sequence *converges* if for every pair (i, j) there exists a number $n = n_{i,j}$ such the two maps

$$\xrightarrow{d_r} \mathbf{E}_r^{i,j} \xrightarrow{d_r}$$

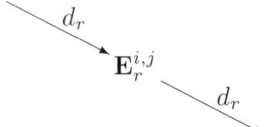

are trivial for all $r \geq n$. The resulting module $\mathbf{E}_n^{i,j} \cong \mathbf{E}_{n+1}^{i,j} \cong \mathbf{E}_{n+2}^{i,j} \cong \mathbf{E}_{n+3}^{i,j} \cong \cdots$ is denoted by $\mathbf{E}_\infty^{i,j}$, and one says that the spectral sequence *converges* to the bigraded module $\mathbf{E}_\infty^{\bullet\bullet}$. The spectral sequence *collapses* if all differentials d_r vanish, and it collapses at $\mathbf{E}_n$ if all differentials vanish for $r \geq n$, in which case $\mathbf{E}_n \cong \mathbf{E}_\infty$. Note also that if $\mathbf{E}_n^{i,j} = 0$, then $\mathbf{E}_r^{i,j} = 0$ for all $r \geq n$.

Each term $\mathbf{E}_r$ can be visualized as the grid $\mathbb{Z}^2 \subseteq \mathbb{R}^2$. The point with coordinates (i, j) represents $\mathbf{E}_r^{i,j}$, and the differentials are arrows between elements of the grid pointing r steps to the right- and $r-1$ steps downwards. As r increases, the arrows become longer and their slope approaches -1.

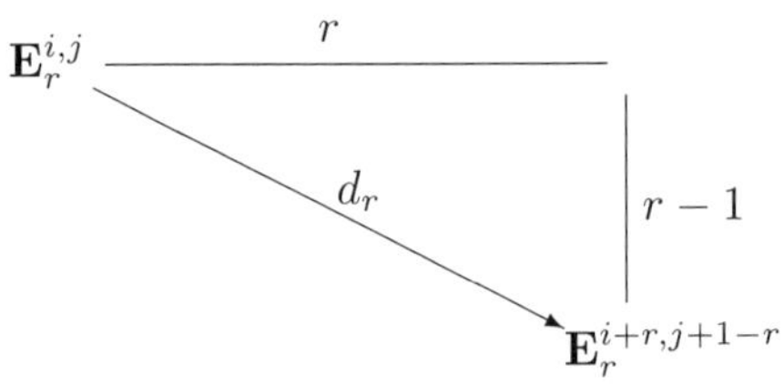

Typically, the $\mathbf{E}_2$-term contains lots of zeros. The first differentials are zero until the arrows become long enough to reach from one non-zero entry to another one. If the region in $\mathbb{R}^2$ containing the non-zero terms is bounded, the arrows become too long after some time, and the spectral sequence collapses.

LEMMA 1.6. *If R is a field, and if* $\dim(\mathbf{E}_n) < \infty$, *then*

$$\dim(\mathbf{E}_\infty) \leq \dim(\mathbf{E}_n).$$

Equality holds if and only if the spectral sequence collapses at n.

PROOF. We have

$$\dim(\mathbf{E}_r) = \dim(\operatorname{im}(d_r)) + \dim(\ker(d_r)),$$

whence

$$\dim(\mathbf{E}_{r+1}) = \dim(\mathbf{E}_r) - 2\dim(\operatorname{im}(d_r)).$$

□

THEOREM 1.7 (The Leray-Serre spectral sequence).

Let

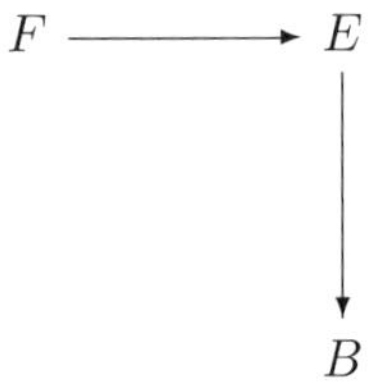

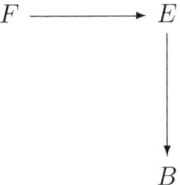

be a fibration over a path-connected space B. The fundamental group $\pi_1(B)$ acts on the fibre F and hence on the cohomology $\mathbf{H}^\bullet(F; R)$. If this action is trivial (e.g. if B is 1-connected), then the fibration is called R-simple. *Suppose that this is the case. Then there is an $\mathbf{E}_2$-spectral sequence which converges to the bigraded module associated to some (bounded and convergent) filtration of $\mathbf{H}^\bullet(E; R)$, with $\mathbf{E}_2^{i,j} \cong \mathbf{H}^i(B; \mathbf{H}^j(F; R))$.*

PROOF. See Spanier [**90**] 9.4.9. □

Note that $\mathbf{E}_r^{i,j} = 0$ if $i < 0$ or $j < 0$. Such a *first-quadrant spectral sequence* is always convergent because the arrows eventually stick out of the first quadrant.

Put $\mathbf{F}^i\mathbf{H}^j = \mathbf{F}^i\mathbf{H}^j(E;R)$. There are short exact sequences

$$0 \longrightarrow \mathbf{F}^i\mathbf{H}^i \xrightarrow{\cong} \mathbf{E}_\infty^{i,0} \longrightarrow 0$$

$$0 \longrightarrow \mathbf{F}^i\mathbf{H}^i \longrightarrow \mathbf{F}^{i-1}\mathbf{H}^i \longrightarrow \mathbf{E}_\infty^{i-1,1} \longrightarrow 0$$

$$0 \longrightarrow \mathbf{F}^{i-1}\mathbf{H}^i \longrightarrow \mathbf{F}^{i-2}\mathbf{H}^i \longrightarrow \mathbf{E}_\infty^{i-2,2} \longrightarrow 0$$

$$\vdots$$

$$0 \longrightarrow \mathbf{F}^1\mathbf{H}^i \longrightarrow \mathbf{H}^i \longrightarrow \mathbf{E}_\infty^{0,i} \longrightarrow 0$$

We consider the 'edges' of the first quadrant, i.e. the X- and Y-axis. Note that

$$\mathbf{E}_\infty^{0,i} \cong \frac{\mathbf{F}^0\mathbf{H}^i}{\mathbf{F}^1\mathbf{H}^i}$$

is a quotient of $\mathbf{F}^0\mathbf{H}^i = \mathbf{H}^i(E;R)$, and that

$$\mathbf{E}_\infty^{0,i} \subseteq \mathbf{E}_2^{0,i}$$

because all arrows which come from the left are zero. Since the base B is assumed to be path connected, $\mathbf{H}^0(B;\mathbf{H}^\bullet(F;R)) \cong \mathbf{H}^\bullet(F;R)$ and the following diagram commutes, cp. Spanier [**90**] 9.5.

$$\begin{array}{ccc}
\mathbf{H}^i(F;R) & \longleftarrow & \mathbf{H}^i(E;R) \\
\cong\big\downarrow & & \big\downarrow \\
\mathbf{E}_2^{0,i} & \hookleftarrow & \mathbf{E}_\infty^{0,i}
\end{array}$$

The projection $E \longrightarrow B$ can also be interpreted in terms of the Leray-Serre spectral sequence. Note that there is a surjection

$$\mathbf{E}_2^{i,0} \longrightarrow \mathbf{E}_\infty^{i,0}$$

because all arrows starting on the X-axis are zero. Moreover,

$$\mathbf{E}_\infty^{i,0} \cong \frac{\mathbf{F}^i\mathbf{H}^i}{\mathbf{F}^{i+1}\mathbf{H}^i} = \mathbf{F}^i\mathbf{H}^i \subseteq \mathbf{H}^i.$$

If the fibre F is path connected, then $\mathbf{H}^\bullet(B;\mathbf{H}^0(F;R)) \cong \mathbf{H}^\bullet(B;R)$ and the diagram

$$\begin{array}{ccc}
\mathbf{H}^i(E;R) & \longleftarrow & \mathbf{H}^i(B;R) \\
\big\uparrow & & \big\uparrow\cong \\
\mathbf{E}_\infty^{i,0} & \longleftarrow & \mathbf{E}_2^{i,0}
\end{array}$$

commutes, cp. Spanier [**90**] 9.5.

1.8. EXAMPLE Here is a very simple example. Suppose that the fibre F is R-acyclic, $\widetilde{\mathbf{H}}^\bullet(F;R) = 0$. Then all non-zero terms of the spectral sequence are contained in the X-axis, the spectral sequence collapses, and $\mathbf{H}^\bullet(E;R) \longleftarrow \mathbf{H}^\bullet(B;R)$ is an isomorphism. A similar result holds if B is R-acyclic; in this case $\mathbf{H}^\bullet(F;R) \xleftarrow{\cong} \mathbf{H}^\bullet(E;R)$.

Note that 1.7 says nothing about the differentials of the spectral sequence. Thus, it is almost impossible to determine $\mathbf{E}_\infty$ from $\mathbf{E}_2$, except for very special situations. An important additional ingredient is the multiplicative structure of the cohomology rings.

1.B. Multiplicative structure

1.9. GRADED ALGEBRAS

A *graded R-algebra* is a graded R-module $M^\bullet$ with an R-linear map

$$\operatorname{Tot}(M \otimes M) \longrightarrow M$$

of degree 0. Thus, M is an R-algebra, with the extra condition that $M^i \cdot M^j \subseteq M^{i+j}$. If there is a unit element 1, then we require that $1 \in M^0$. The multiplication is *graded commutative* if

$$m_1 \cdot m_2 = (-1)^{\deg(m_1)\deg(m_2)} m_2 \cdot m_1$$

holds for all homogeneous elements. If M, N are graded algebras, then $\operatorname{Tot}(M \otimes N)$ is again a graded algebra if we define

$$(m_1 \otimes n_1) \cdot (m_2 \otimes n_2) = (-1)^{\deg(n_1)\deg(m_2)} (m_1 \cdot m_2) \otimes (n_1 \cdot n_2).$$

If M, N are graded commutative, then so is their tensor product. In the case of a filtration of a graded algebra M we require that

$$\mathbf{F}^i M^j \cdot \mathbf{F}^k M^l \subseteq \mathbf{F}^{i+k} M^{j+l}.$$

The corresponding notions in the bigraded case are very similar. Here, we require that

$$M^{i,j} \cdot M^{k,l} \subseteq M^{i+k,j+l};$$

the multiplication is *bigraded commutative* if it is graded commutative with respect to the total degree.

1.10. DIFFERENTIAL GRADED ALGEBRAS

If d is a differential on a graded algebra M, and if d satisfies the Leibniz rule

$$d(m_1 \cdot m_2) = d(m_1) \cdot m_2 + (-1)^{\deg(m_1)} m_1 \cdot d(m_2),$$

then (M, d) is called a *differential graded algebra*. Similarly, a *differential bigraded algebra* is a bigraded algebra M with a differential of bidegree $(r, 1-r)$, such that $\operatorname{Tot}(M)$ is a differential graded algebra, i.e.

$$d(m_1 \cdot m_2) = d(m_1) \cdot m_2 + (-1)^{i+j} m_1 \cdot d(m_2),$$

where $\deg(m_1) = (i, j)$.

Now we can sharpen the statement about the Leray-Serre spectral sequence. The $\mathbf{E}_2$-term $\mathbf{E}_2^{\bullet\bullet} \cong \mathbf{H}^\bullet(B; \mathbf{H}^\bullet(F; R))$ is a differential bigraded algebra. If R is a field, then there is an isomorphism of (bi)graded algebras

$$\mathbf{H}^\bullet(B; \mathbf{H}^\bullet(F; R)) \cong \mathbf{H}^\bullet(B; R) \otimes \mathbf{H}^\bullet(F; R).$$

Note that the multiplication on the right-hand side is

$$(b_1 \otimes f_1) \cdot (b_2 \otimes f_2) = (-1)^{\deg(f_1)\deg(b_2)}(b_1 \smile b_2) \otimes (f_1 \smile f_2).$$

The cohomology of a differential graded algebra is again a graded algebra. The differentials in the Leray-Serre spectral sequence are compatible with this algebra structure, i.e. the $\mathbf{E}_r$ are differential bigraded algebras, and $\mathbf{E}_\infty$ is a graded algebra. Moreover, the corresponding filtration on $\mathbf{H}^\bullet(E; R)$ is compatible with the cup-product.

The following examples illustrate the multiplicative properties; they will also be useful later.

LEMMA 1.11. *Suppose that $\mathbf{H}^\bullet(F; \mathbb{Q}) \cong \bigwedge_{\mathbb{Q}}(u)$ is an exterior algebra on one homogeneous generator u of odd degree n. Suppose that*

$$\begin{array}{ccc} F & \longrightarrow & E \\ & & \big\downarrow \\ & & B \end{array}$$

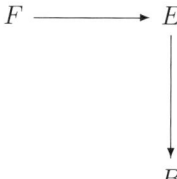

is a $\mathbb{Q}$-simple fibration, and that E is $\mathbb{Q}$-acyclic. Then $\mathbf{H}^\bullet(B; \mathbb{Q}) \cong \mathbb{Q}[a]$ is a polynomial algebra on one generator a of degree $n+1$.

PROOF. The only non-zero terms in $\mathbf{E}_2$ (and hence in $\mathbf{E}_r$, $2 \le r \le \infty$) are contained in the two horizontal strips $\mathbf{E}_2^{\bullet,0}$ and $\mathbf{E}_2^{\bullet,n}$. Thus, the only possibly non-zero differential is d_{n+1}, and $\mathbf{E}_{n+2} \cong \mathbf{E}_\infty$. Therefore, the sequences

$$0 \longrightarrow \mathbf{E}_{n+1}^{i,n} \longrightarrow \mathbf{E}_{n+1}^{i+n+1,0} \longrightarrow 0$$

are exact for all $i \in \mathbb{Z}$. But

$$\mathbf{E}_{n+1}^{i,n} \cong \mathbf{E}_2^{i,n} \cong \mathbf{H}^i(B; \mathbb{Q})$$

and

$$\mathbf{E}_{n+1}^{i+n+1,0} \cong \mathbf{E}_2^{i+n+1,0} \cong \mathbf{H}^{i+n+1}(B; \mathbb{Q}).$$

This shows already that

$$\mathbf{H}^i(B; \mathbb{Q}) \cong \begin{cases} \mathbb{Q} & \text{for } i \equiv 0 \pmod{n+1} \\ 0 & \text{else.} \end{cases}$$

Let $d_{n+1}(1 \otimes u) = a \otimes 1$, for some $a \in \mathbf{H}^{n+1}(B; \mathbb{Q})$. We claim that a^k spans $\mathbf{H}^{k(n+1)}(B; \mathbb{Q})$, for all $k \ge 0$. This is true for $k = 1$, so we proceed by induction. Suppose that a^k spans $\mathbf{H}^{k(n+1)}(B; \mathbb{Q})$. Then $a^k \otimes u$ spans $\mathbf{E}_2^{k(n+1),n} \cong \mathbf{E}_{n+1}^{k(n+1),n}$. Now

$$d_{n+1}(a^k \otimes u) = d_{n+1}((a^k \otimes 1) \cdot (1 \otimes u)) = (a^k \otimes 1) \cdot (a \otimes 1) = a^{k+1} \otimes 1.$$

This element spans $\mathbf{E}_{n+1}^{(k+1)(n+1),0}$, and the claim follows. □

Suppose now that $\mathbf{H}^\bullet(F;\mathbb{Q})$ is a polynomial algebra in an element a of even degree. Here, the multiplicative structure is more important.

LEMMA 1.12. *Suppose that $\mathbf{H}^\bullet(F;\mathbb{Q}) \cong \mathbb{Q}[a]$ is a polynomial algebra in a homogeneous generator a of even degree n. Suppose that*

$$\begin{array}{ccc} F & \longrightarrow & E \\ & & \downarrow \\ & & B \end{array}$$

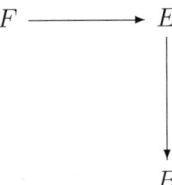

is a $\mathbb{Q}$-simple fibration, and that E is $\mathbb{Q}$-acyclic. Then $\mathbf{H}^\bullet(B;\mathbb{Q}) \cong \bigwedge_{\mathbb{Q}}(u)$ is an exterior algebra on one generator u of degree $n+1$.

PROOF. The non-zero terms of $\mathbf{E}_2 \cong \mathbf{H}^\bullet(B;\mathbb{Q}) \otimes \mathbb{Q}[a]$ are contained in the horizontal strips $\mathbf{E}_2^{\bullet,kn}$, for $k \geq 0$. Thus, the differentials $d_2, \ldots d_n$ vanish and $\mathbf{E}_{n+1} \cong \mathbf{E}_2$. Now $\mathbf{E}_2^{i,0} \cong \mathbf{H}^i(B;\mathbb{Q}) = 0$ for $1 < i < n+1$, and the sequence

$$0 \longrightarrow \mathbf{E}_{n+1}^{0,n} \xrightarrow{d_{n+1}} \mathbf{E}_{n+1}^{n+1,0} \longrightarrow 0$$

is exact. Put $d_{n+1}(1 \otimes a) = u \otimes 1$. By the Leibniz rule, $d_{n+1}(1 \otimes a^k) = ku \otimes a^{k-1}$, whence

$$d_{n+1}(u \otimes a^k) = 0$$

for $k \geq 1$. The bigraded subalgebra $\bigwedge_{\mathbb{Q}}(u) \otimes \mathbb{Q}[a] \subseteq \mathbf{E}_{n+1}$ generated by a and u is thus closed under the differential d_{n+1}, and its cohomology is trivial, as is easily seen (using the divisibility of $\mathbb{Q}$). Suppose $\mathbf{H}^\bullet(B;\mathbb{Q}) \neq \bigwedge_{\mathbb{Q}}(u)$. Then there is a minimal number $i \geq 1$ such that $\mathbf{H}^i(B;\mathbb{Q}) \setminus \bigwedge_{\mathbb{Q}}(u) \neq 0$. But now $\mathbf{E}_{n+2}^{j,k} = 0$ for $0 < j < i$ and $k \geq 0$. Thus, all arrows that start or end at $\mathbf{E}_{n+2}^{i,0} \cong \mathbf{H}^i(B;\mathbb{Q})$ are zero. This contradicts the fact that $\mathbf{E}_\infty^{i,0} = 0$. □

The two lemmata have an immediate application. Let π be an abelian group. Recall that an Eilenberg-MacLane space of type (π, n) is a space $K(\pi, n)$ with the property that

$$\pi_k(K(\pi,n)) \cong \begin{cases} \pi & \text{for } k = n \\ 0 & \text{else.} \end{cases}$$

Let $PK(\pi,n)$ denote the path space. The map that sends a path to its endpoint is a fibration whose fibre (over the base point) is the loop space $\Omega K(\pi,n)$,

$$\begin{array}{ccc} \Omega K(\pi,n) & \longrightarrow & PK(\pi,n) \\ & & \downarrow \\ & & K(\pi,n). \end{array}$$

The path space is contractible and hence acyclic. The long exact homotopy sequence of this fibration shows then that $\Omega K(\pi,n)$ is a space of type $K(\pi, n-1)$. Now $\mathbb{S}^1$ is clearly a space of type $(\mathbb{Z}, 1)$. Combining the two lemmata we see the following.

PROPOSITION 1.13. *Let $K(\mathbb{Z}, n)$ be a space of type $(\mathbb{Z}, n)$. If n is odd, then the rational cohomology*

$$\mathbf{H}^\bullet(K(\mathbb{Z}, n); \mathbb{Q}) \cong \bigwedge(u)$$

is an exterior algebra on one homogeneous generator of degree n. If n is even, then

$$\mathbf{H}^\bullet(K(\mathbb{Z}, n); \mathbb{Q}) \cong \mathbb{Q}[a]$$

is a polynomial algebra on one homogeneous generator a of degree n.

PROOF. The 1-sphere $\mathbb{S}^1$ is clearly a space of type $(\mathbb{Z}, 1)$. The proof is a straight-forward induction based on the two lemmata and the path-space fibration. □

If π is a finite cyclic group, then a space of type (π, n) is $\mathbb{Q}$-acyclic, cp. Spanier [**90**] 9.5.6. This, combined with the result above, yields the following.

PROPOSITION 1.14. *Let π be a finitely generated abelian group of rank k. The rational cohomology of a space of type (π, n) is a free graded commutative algebra on k homogeneous generators of degree n. Thus, if n is odd, then*

$$\mathbf{H}^\bullet(K(\pi, n); \mathbb{Q}) \cong \bigwedge(u_1, \ldots, u_k),$$

where $\deg(u_1) = \ldots = \deg(u_k) = n$, *and if n is even then*

$$\mathbf{H}^\bullet(K(\pi, n); \mathbb{Q}) \cong \mathbb{Q}[a_1, \ldots, a_k],$$

where $\deg(a_1) = \ldots = \deg(a_k) = n$.

PROOF. Let $\pi = C_1 \oplus \cdots \oplus C_r$ be a direct sum of cyclic groups. Any two Eilenberg-MacLane spaces of type (π, n) are (weakly) homotopy equivalent; thus

$$K(\pi, n) \simeq K(C_1, n) \times \cdots \times K(C_r, n),$$

and the claim follows from the Künneth Theorem and 1.13. □

1.C. Notes on collapsing

We collect some criteria which ensure that the Leray-Serre spectral sequence collapses.

1.15. Let

$$\begin{array}{ccc} F & \xrightarrow{\iota} & E \\ & & \downarrow{\scriptstyle p} \\ & & B \end{array}$$

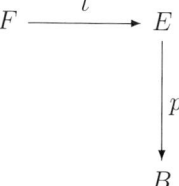

be an R-simple fibration. Assume that either $\mathbf{H}^\bullet(F; R)$ or $\mathbf{H}^\bullet(B; R)$ is a free R-module. Then the $\mathbf{E}_2$-term of the Leray-Serre spectral sequence is

$$\mathbf{E}_2 \cong \mathbf{H}^\bullet(B; R) \otimes \mathbf{H}^\bullet(F; R).$$

Suppose that the first differentials $d_2, d_3, \dots, d_{k-1}$ vanish, $\mathbf{E}_2 \cong \mathbf{E}_3 \cong \dots \cong \mathbf{E}_k$. Because of the Leibniz rule, the differential d_k is completely determined by its restriction to the Y-axis $\mathbf{E}_k^{0,\bullet}$:

$$\begin{aligned} d_k(b \otimes f) &= d_k((b \otimes 1) \cdot (1 \otimes f)) \\ &= \underbrace{d_k(b \otimes 1)}_{0} \cdot (1 \otimes f) + (-1)^{\deg(b)}(b \otimes 1) \cdot d_k(1 \otimes f) \\ &= (-1)^{\deg(b)}(b \otimes 1) \cdot d_k(1 \otimes f) \end{aligned}$$

In particular, if d_k is trivial on $\mathbf{E}_k^{0,\bullet}$, then $d_k = 0$.

DEFINITION 1.16. If the map

$$\iota^\bullet : \mathbf{H}^\bullet(E; R) \longrightarrow \mathbf{H}^\bullet(F; R)$$

is surjective, then F is called *totally non-homologous to 0* in E.

THEOREM 1.17 (Leray-Hirsch). *Suppose that*

$$\begin{array}{ccc} F & \xrightarrow{\ \iota\ } & E \\ & & \Big\downarrow p \\ & & B \end{array}$$

is an R-simple fibration, and that F and B are of finite type. Suppose moreover that $\mathbf{H}^\bullet(F)$ is a finitely generated free R-module. If F is totally non-homologous to 0, then the Leray-Serre spectral sequence collapses, $p^\bullet$ is an injection, and there is an additive isomorphism

$$\mathbf{H}^\bullet(E) \cong \mathbf{H}^\bullet(B) \otimes \mathbf{H}^\bullet(F).$$

The kernel of the map $\mathbf{H}^\bullet(F; R) \longleftarrow \mathbf{H}^\bullet(E; R)$ is the ideal generated by

$$\operatorname{im}(p^\bullet) \cap \bigoplus_{i \geq 1} \mathbf{H}^i(E; R).$$

PROOF. See Spanier [**90**] 5.7.9 and Mimura-Toda [**71**] III Thm. 4.2. □

CHAPTER 2

Ranks of homotopy groups

Suppose that X is a 1-connected space which has the same integral (co-)homology as a product of spheres. The aim of this chapter is to determine the ranks of the homotopy groups of X in terms of the rational cohomology of X. The idea is to generalize a result of Cartan-Serre [**20**] who determined the rational homotopy groups of spaces whose rational cohomology ring is freely generated. The note of Cartan and Serre appeared in the Comptes Rendus, and they gave only a short hint how to prove their theorem; a proof can be found in Fomenko-Fuchs-Gutenmacher [**35**]. We prove here a more general version of the Cartan-Serre Theorem.

Everything in this chapter could also be done in the framework of rational homotopy theory and minimal models, see e.g. Félix-Halperin-Thomas [**33**]. I chose a more traditional homotopy theoretic approach.

2.A. The Whitehead tower

Let X be a 1-connected space, and let $\pi_k = \pi_k(X)$ be the kth homotopy group of X. The idea is the following. There is a sequence of spaces and fibrations

$$\cdots \longrightarrow X\langle k+1\rangle \longrightarrow X\langle k\rangle \longrightarrow X\langle k-1\rangle \longrightarrow \cdots \longrightarrow X\langle 1\rangle = X$$

with the following properties.

WT$_1$ Each $X\langle k\rangle$ is $(k-1)$-connected.

WT$_2$ The map $X\langle k\rangle \longrightarrow X$ induces an isomorphism on all homotopy groups of degree at least k.

WT$_3$ The homotopy fibre of the map $X\langle k+1\rangle \longrightarrow X\langle k\rangle$ is an Eilenberg-MacLane space of type $(\pi_k, k-1)$.

This is sometimes called the *Whitehead tower* or *upside-down Postnikov tower* of X. It is rather easy to compute the rank of π_k in terms of the cohomology group $\mathbf{H}^k(X\langle k\rangle;\mathbb{Q})$:

$$\mathrm{rk}(\pi_k) = \mathrm{rk}(\pi_k(X\langle k\rangle)) = \dim_{\mathbb{Q}} \mathbf{H}^k(X\langle k\rangle;\mathbb{Q}).$$

Thus, if there is a way to calculate the rational cohomology of the spaces $X\langle k\rangle$ from the knowledge of $\mathbf{H}^\bullet(X;\mathbb{Q})$, then one can determine the ranks of the homotopy groups.

The Whitehead tower can be constructed as follows. Suppose we have already constructed $X\langle k\rangle$. Then $X\langle k\rangle$ is $(k-1)$-connected. There exists an Eilenberg-MacLane space $K(\pi_k, k)$ and a map $\theta : X\langle k\rangle \longrightarrow K(\pi_k, k)$ that induces an isomorphism between $\pi_k(X\langle k\rangle)$ and $\pi_k(K(\pi_k, k))$ (one way to obtain this space is to kill all higher-dimensional homotopy groups of $X\langle k\rangle$, cp. e.g. Whitehead [**113**] V.2.4).

Consider the path space fibration

$$\begin{array}{ccc} \Omega K(\pi_k, k) & \longrightarrow & PK(\pi_k, k) \\ & & \big\downarrow \\ & & K(\pi_k, k). \end{array}$$

Since the path space of any 0-connected space is contractible, $\Omega K(\pi, k)$ is an Eilenberg-MacLane space of type $(\pi, k-1)$. We pull the path-space fibration $PK(\pi_k, k) \longrightarrow K(\pi_k, k)$ back via θ and denote the resulting total space by $X\langle k+1\rangle$,

$$\begin{array}{ccccc} \Omega K(\pi_k, k) & \longrightarrow & X\langle k+1\rangle & \longrightarrow & PK(\pi_k, k) \\ & & \big\downarrow & & \big\downarrow \\ & & X\langle k\rangle & \xrightarrow{\ \theta\ } & K(\pi_k, k). \end{array}$$

The interesting part of the exact homotopy sequences of these fibrations is

$$\begin{array}{ccccccc} \longrightarrow & \pi_k(K(\pi_k, k)) & \underset{\cong}{\xrightarrow{\ \partial\ }} & \pi_{k-1}(K(\pi_k, k-1)) & \longrightarrow \\ & \big\downarrow{\scriptstyle\cong} & & \big\| & \\ \longrightarrow & \pi_k(X\langle k\rangle) & \xrightarrow{\ \partial\ } & \pi_{k-1}(K(\pi_k, k-1)) & \longrightarrow & . \end{array}$$

It shows that $\pi_k(X\langle k+1\rangle) \cong 0 \cong \pi_{k-1}(X\langle k+1\rangle)$; thus, $X\langle k+1\rangle$ has the claimed properties $\mathbf{WT}_1$, $\mathbf{WT}_2$, **and** $\mathbf{WT}_3$.

Suppose that π_k is *finite*. Then $K(\pi_k, k-1)$ is $\mathbb{Q}$-acyclic, and $\mathbf{H}^\bullet(X\langle k+1\rangle; \mathbb{Q}) \xleftarrow{\cong} \mathbf{H}^\bullet(X\langle k\rangle; \mathbb{Q})$ is an isomorphism, cp. 1.8. So, finite homotopy groups in low degrees do not matter.

LEMMA 2.1. *Let π_k be the first infinite homotopy group of X. Then there is a chain of isomorphisms*

$$\mathbf{H}^\bullet(X\langle k\rangle; \mathbb{Q}) \xleftarrow{\cong} \mathbf{H}^\bullet(X\langle k-1\rangle; \mathbb{Q}) \xleftarrow{\cong} \mathbf{H}^\bullet(X\langle k-2\rangle; \mathbb{Q}) \xleftarrow{\cong} \cdots \xleftarrow{\cong} \mathbf{H}^\bullet(X; \mathbb{Q}).$$

□

By Hurewicz' Theorem we have $\pi_k \cong \mathbf{H}_k(X\langle k\rangle)$. Since

$$\mathbf{H}^k(X\langle k\rangle; \mathbb{Q}) \cong \mathrm{Hom}(\mathbf{H}_k(X\langle k\rangle), \mathbb{Q}) \cong \mathrm{Hom}(\pi_k, \mathbb{Q}) \cong \mathbb{Q}^{\mathrm{rk}(\pi_k)},$$

we have the following result.

LEMMA 2.2. *Let $\mathbf{H}^k(X; \mathbb{Q})$ be the first non-trivial rational cohomology group of X, for $k \geq 1$. Then*

$$\mathrm{rk}(\pi_k) = \dim_{\mathbb{Q}} \mathbf{H}^k(X; \mathbb{Q}),$$

and all groups π_j are finite, for $0 \leq j < k$ (note that we assumed that $\pi_0 = \pi_1 = 0$). □

So one can tell the rank of the first infinite homotopy group right away from the rational cohomology of X. Suppose now that $\mathrm{rk}(\pi_k) = r$. The question is if we can compute the cohomology of $X\langle k+1\rangle$. In the diagram

$$\begin{array}{ccccc} \Omega K(\pi_k, k) & \longrightarrow & X\langle k+1\rangle & \longrightarrow & PK(\pi_k, k) \\ & & \downarrow & & \downarrow \\ & & X\langle k\rangle & \xrightarrow{\theta} & K(\pi_k, k), \end{array}$$

we consider first the fibration on the right

$$\begin{array}{ccc} \Omega K(\pi_k, k) & \longrightarrow & PK(\pi_k, k) \\ & & \downarrow \\ & & K(\pi_k, k). \end{array}$$

This is very similar to what we did in 1.11, 1.12. Recall that $\Omega K(\pi_k, k)$ is a space of type $(\pi_k, k-1)$,

$$\Omega K(\pi_k, k) \simeq K(\pi_k, k-1).$$

Consider the Leray-Serre spectral sequence of this fibration. The total space $PK(\pi_k, k)$ is acyclic, hence the $\mathbf{E}_\infty$-term is trivial. Since we are working over the field $\mathbb{Q}$, the $\mathbf{E}_2$-term is given by

$$\mathbf{E}_2^{s,t} \cong \mathbf{H}^s(K(\pi, k); \mathbb{Q}) \otimes \mathbf{H}^t(K(\pi, k-1); \mathbb{Q}).$$

Let $k \geq 2$ be *even*. Then $\mathbf{H}^\bullet(K(\pi, k-1); \mathbb{Q})$ is an exterior algebra. Thus, the only non-trivial terms in $\mathbf{E}_2$ (and hence in $\mathbf{E}_n$, $2 \leq n \leq \infty$) are contained in the horizontal strips $\mathbf{E}_2^{\bullet,0}$, $\mathbf{E}_2^{\bullet,k-1}, \ldots, \mathbf{E}_2^{\bullet,r(k-1)}$. Therefore, the first possibly non-trivial differential is d_k, and

$$\mathbf{E}_2 \cong \mathbf{E}_3 \cong \cdots \cong \mathbf{E}_k.$$

The $\mathbf{E}_k$-term looks as follows.

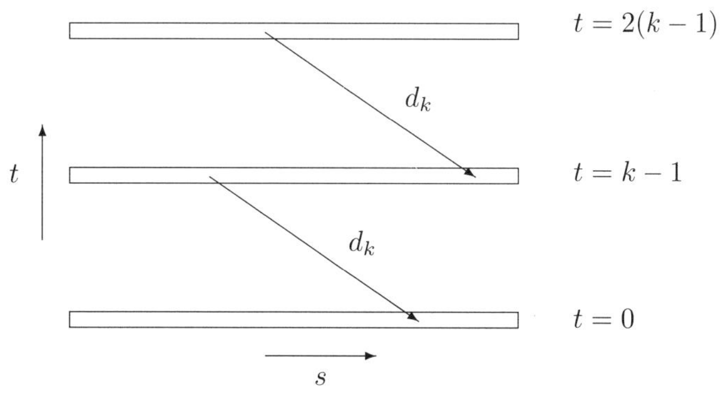

Since $\mathbf{E}_\infty^{0,k-1} = 0$, the sequence

$$0 \longrightarrow \mathbf{E}_n^{0,k-1} \xrightarrow{d_k} \mathbf{E}_k^{k,0} \longrightarrow 0$$

is exact. Let $w_1, \dots, w_r$ be a basis for $\mathbf{H}^{k-1}(K(\pi, k-1); \mathbb{Q})$. We define a $\mathbb{Q}$-linear map τ by

$$d_k(1 \otimes w_i) = \tau(w_i) \otimes 1.$$

The elements $\tau(w_1), \dots, \tau(w_r)$ span $\mathbf{H}^k(K(\pi, k); \mathbb{Q})$, therefore, $\mathbf{H}^\bullet(K(\pi, k); \mathbb{Q})$ is a (free graded) polynomial algebra in these elements,

$$\mathbf{H}^\bullet(K(\pi, k); \mathbb{Q}) = \mathbb{Q}[b_1, \dots, b_r] = \mathbb{Q}[\tau(w_1), \dots, \tau(w_r)].$$

We claim that $\mathbf{E}_{k+1}$ is trivial. Put

$$\mathbf{C}_i^{\bullet\bullet} = \mathbb{Q}[\tau(w_i)] \otimes \bigwedge\nolimits_{\mathbb{Q}}(w_i).$$

This is a bigraded subalgebra of $\mathbf{E}_k$ which is closed under the differential d_k, and it is easy to check that the cohomology of each $\mathbf{C}_i$ is trivial, $\mathbf{H}(\mathbf{C}_i) = 0$. But

$$\mathrm{Tot}(\mathbf{E}_k) = \mathrm{Tot}(\mathbf{C}_1 \otimes \cdots \otimes \mathbf{C}_r).$$

By the Künneth Theorem,

$$\mathbf{H}(\mathbf{E}_k) \cong \mathbf{H}(\mathbf{C}_1) \otimes \cdots \otimes \mathbf{H}(\mathbf{C}_r)$$

is trivial, and thus $\mathbf{E}_{k+1} = \mathbf{E}_{k+2} = \cdots = \mathbf{E}_\infty$.

If k is *odd*, then the non-trivial terms in $\mathbf{E}_2$ are in the vertical strips $\mathbf{E}_2^{0,\bullet}$, $\mathbf{E}_2^{k,\bullet}, \dots, \mathbf{E}_2^{kr,\bullet}$. Again,

$$\mathbf{E}_2 \cong \mathbf{E}_3 \cong \cdots \cong \mathbf{E}_k$$

and the first possibly non-trivial differential is d_k.

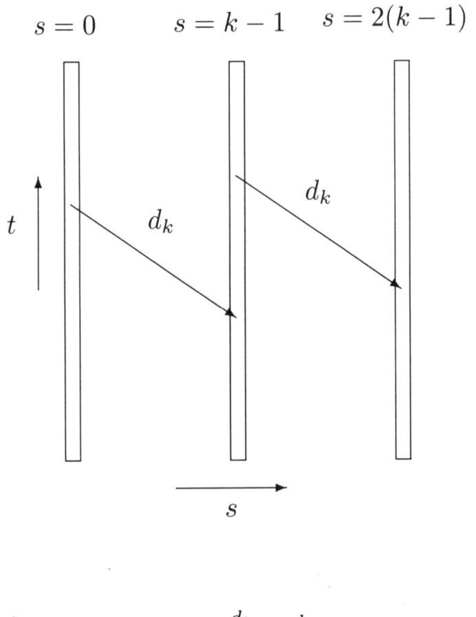

As before the sequence

$$0 \longrightarrow 1 \otimes \mathbf{H}^{k-1}(K(\pi, k-1); \mathbb{Q}) \xrightarrow{d_k} \mathbf{H}^k(K(\pi, k); \mathbb{Q}) \otimes 1 \longrightarrow 0$$

is exact, and we may put

$$d_k(1 \otimes b_i) = \tau(b_i) \otimes 1.$$

It follows that

$$\mathbf{H}^\bullet(K(\pi, k); \mathbb{Q}) = \bigwedge\nolimits_{\mathbb{Q}}(w_1, \dots, w_r) = \bigwedge\nolimits_{\mathbb{Q}}(\tau(b_1), \dots, \tau(b_r))$$

is an exterior algebra generated by the $\tau(b_i)$. By the same reasoning as above, $\mathbf{E}_{k+1} \cong \mathbf{E}_{k+2} \cong \cdots \cong \mathbf{E}_\infty$ is trivial.

2.B. The Cartan-Serre Theorem

Now we get back to our real problem, the fibration

$$\begin{array}{ccc} \Omega K(\pi_k, k) & \longrightarrow & X\langle k+1\rangle & & \\ & & \downarrow & & \\ & & X\langle k\rangle & \xrightarrow{\theta} & K(\pi_k, k). \end{array}$$

The map in cohomology $\theta^\bullet$ induces a map between the two spectral sequences. The image of $\theta^\bullet$ is precisely the subalgebra of $\mathbf{H}^\bullet(X\langle k\rangle;\mathbb{Q})$ generated by $\mathbf{H}^k(X\langle k\rangle;\mathbb{Q})$.

Suppose that k is even. Again, the non-zero terms in $\mathbf{E}_2$ are contained in the horizontal strips $\mathbf{E}_2^{\bullet,j(k-1)}$, for $0 \leq j \leq r$, and the same is true for $\mathbf{E}_i$, $i \geq 2$. Therefore, the first possibly non-trivial differential again is d_k, and

$$\mathbf{E}_2 \cong \mathbf{E}_3 \cong \cdots \cong \mathbf{E}_k.$$

Moreover, we know d_k on the image of $\theta^\bullet$. Let $\mathbf{A}^\bullet = \theta^\bullet(\mathbf{H}^\bullet(K(\pi_k, k);\mathbb{Q}))$ denote this image. Suppose that the following holds: the cohomology algebra $\mathbf{H}^\bullet(X\langle k\rangle;\mathbb{Q})$ decomposes as

$$\mathbf{H}^\bullet(X\langle k\rangle;\mathbb{Q}) \cong \mathrm{Tot}(\mathbf{A}^\bullet \otimes \mathbf{B}^\bullet),$$

for some subalgebra $\mathbf{B}^\bullet \subseteq \mathbf{H}^\bullet(X\langle k\rangle;\mathbb{Q})$. This holds for example if $\mathbf{H}^\bullet(X\langle k\rangle;\mathbb{Q})$ is a free graded commutative algebra. Endow the complex $\mathbf{B}^\bullet$ with the trivial (zero) differential. Then there is an isomorphism of differential graded algebras

$$\mathbf{E}_k^{\bullet\bullet} \cong \mathrm{Tot}(\mathbf{A}^\bullet \otimes \mathbf{H}^\bullet(K(\pi, k-1);\mathbb{Q})) \otimes \mathbf{B}^\bullet,$$

and the Künneth Theorem yields

$$\mathrm{Tot}(\mathbf{E}_{k+1}) = \mathrm{Tot}(\mathbf{H}(\mathbf{E}_k^{\bullet\bullet})) = \mathrm{Tot}(\mathbf{H}(\mathbf{A}^\bullet \otimes \mathbf{H}^\bullet(K(\pi, k-1);\mathbb{Q})) \otimes \mathbf{B}^\bullet).$$

If the differential graded algebra $\mathbf{A}^\bullet \otimes \mathbf{H}^\bullet(K(\pi, k-1);\mathbb{Q})$ is acyclic, then we obtain the classical Cartan-Serre Theorem.

THEOREM 2.3 (Cartan-Serre). *Suppose that $\mathbf{H}^\bullet(X;\mathbb{Q})$ is a free graded anti-commutative algebra on homogeneous generators $a_1, \ldots, a_r$ of even degrees and homogeneous generators $u_1, \ldots, u_s$ of odd degrees*

$$\mathbf{H}^\bullet(X;\mathbb{Q}) \cong \mathbb{Q}[a_1, \ldots, a_r] \otimes \bigwedge\nolimits_{\mathbb{Q}}(u_1, \ldots, u_s).$$

Let

$$r_k = |\{i|\ \deg(a_i) = k\}| + |\{i|\ \deg(u_i) = k\}|$$

denote the number of homogeneous generators of degree k. Then

$$\mathrm{rk}(\pi_k(X)) = r_k$$

for all k. In particular, only finitely many homotopy groups of X are infinite, cp. Cartan-Serre **[20]** *Prop. 3.*

PROOF. Let k be the minimum of the degrees of the homogeneous generators. By Lemma 2.1 we have an isomorphism $\mathbf{H}^\bullet(X\langle k\rangle;\mathbb{Q}) \longleftarrow \mathbf{H}^\bullet(X;\mathbb{Q})$, and for $1 \leq i \leq k$ the groups π_i are finite. Suppose that k is even and that $a_1, \dots, a_m$ span $\mathbf{H}^k(X\langle k\rangle;\mathbb{Q})$. By Lemma 2.2 $\mathrm{rk}(\pi_k) = m$. We may put

$$\mathbf{B}^\bullet = \mathbb{Q}[a_{m+1}, \dots, a_r] \otimes \bigwedge\nolimits_{\mathbb{Q}}(u_1, \dots, u_s).$$

Then $\mathbf{H}^\bullet(K(\pi_k, k);\mathbb{Q})$ maps isomorphically onto

$$\mathbf{A}^\bullet = \mathbb{Q}[a_1, \dots, a_m] \subseteq \mathbf{H}^\bullet(X\langle k\rangle;\mathbb{Q}).$$

In the spectral sequence we have

$$\begin{aligned}\mathbf{E}_{k+1}^\bullet &\cong \mathbf{H}([\mathbf{A}^\bullet \otimes \mathbf{H}^\bullet(K(\pi, k-1);\mathbb{Q})] \otimes \mathbf{B}^\bullet) \\ &\cong \mathbf{H}(\mathbf{A}^\bullet \otimes \mathbf{H}^\bullet(K(\pi, k-1);\mathbb{Q})) \otimes \mathbf{H}(\mathbf{B}^\bullet)) \\ &\cong \mathbf{H}(\mathbf{B}^\bullet) \\ &\cong \mathbf{B}^\bullet,\end{aligned}$$

where all non-zero terms are contained in the strip $\mathbf{E}_{k+1}^{\bullet,0}$. Thus,

$$\mathbf{H}^\bullet(X\langle k+1\rangle;\mathbb{Q}) \cong \mathbb{Q}[a_{m+1}, \dots, a_r] \otimes \bigwedge\nolimits_{\mathbb{Q}}(u_1, \dots, u_s)$$

is isomorphic to $\mathbf{H}^\bullet(X\langle k\rangle;\mathbb{Q})$ factored by the ideal generated by $\mathbf{H}^k(X\langle k\rangle;\mathbb{Q})$. The result follows now by induction on the number of generators, starting with the case where X is $\mathbb{Q}$-acyclic. The case when k is odd is similar. □

This implies of course Serre's famous result: the only infinite homotopy group of $\mathbb{S}^{2n+1}$ is π_{2n+1}. However, we need also to consider spaces which have the same cohomology as a product of an odd- and and an even-dimensional sphere. This is not covered by the Cartan-Serre theorem. Thus, suppose that the rational cohomology of X is of the form

$$\mathbf{H}^\bullet(X;\mathbb{Q}) \cong \mathbb{Q}[a]/(a^m) \otimes \bigwedge\nolimits_{\mathbb{Q}}(u_1, \dots, u_r).$$

By the reduction process of the Whitehead tower, there is no loss of generality in assuming that $k = \deg(a) < \deg(u_i)$ for $i = 1, \dots, r$. Then we may put

$$\mathbf{A}^\bullet = \mathbb{Q}[a]/(a^m)$$

and

$$\mathbf{B}^\bullet = \bigwedge\nolimits_{\mathbb{Q}}(u_1, \dots, u_r).$$

Let $d_k(1 \otimes v) = a \otimes 1$. The cohomology of the differential bigraded algebra

$$\mathbb{Q}[a]/(a^m) \otimes \bigwedge\nolimits_{\mathbb{Q}}(v) \subseteq \mathbf{E}_k$$

is $\bigwedge_{\mathbb{Q}}(w)$, for an element w of bidegree $(k(m-1), k-1)$ (and total degree $km-1$). Therefore $\mathbf{E}_{k+1} \cong \bigwedge_{\mathbb{Q}}(w, u_1, \dots, u_r)$. At this stage the spectral sequence collapses, and thus

$$\mathbf{E}_\infty \cong \mathbf{H}^\bullet(X\langle k+1\rangle;\mathbb{Q}) \cong \bigwedge\nolimits_{\mathbb{Q}}(w, u_1, \dots, u_r).$$

THEOREM 2.4. *Let X be a 1-connected space whose rational cohomology is of the form*

$$\mathbf{H}^\bullet(X;\mathbb{Q}) \cong \mathbb{Q}[a]/(a^m) \otimes \bigwedge\nolimits_{\mathbb{Q}}(u_1, \dots, u_r),$$

where $\deg(a)$ *is even and the degrees of the u_i are odd. Let* $r_i = |\{j|\ \deg(u_j) = i\}|$. *Then*

$$\mathrm{rk}(\pi_k) = \begin{cases} 1 & \text{for } k = \deg(a) \\ r_k & \text{for } k \neq \deg(a) \text{ and } k \neq m \cdot \deg(a) - 1 \\ r_k + 1 & \text{for } k = m \cdot \deg(a) - 1. \end{cases}$$

□

In particular, if X has the same cohomology as $\mathbb{S}^{2n}$, i.e. if $\mathbf{H}^\bullet(X;\mathbb{Q}) \cong \mathbb{Q}[a]/(a^2)$, then the only infinite homotopy groups are π_{2n} and π_{4n-1}. This is the even-dimensional version of Serre's finiteness result for homotopy groups of spheres.

CHAPTER 3

Some homogeneous spaces

In this chapter we analyze the algebraic topology of certain homogeneous spaces, using the homotopy theoretic results of the previous chapter. The spaces we are interested in have the same cohomology as a product of two spheres. Although some of the theorems are tailored for this special situation, it should be emphasized that the methods developed here can be adapted to a much larger class of homogeneous spaces.

The first section contains various results about the structure and the topology of compact Lie groups. In the second section we determine the rational Leray-Serre spectral sequence of the principal bundle

$$H \longrightarrow G \longrightarrow G/H$$

where G/H has the same cohomology as $\mathbb{S}^{n_1} \times \mathbb{S}^{n_2}$, where $2 \leq n_1 \leq n_2$ and n_2 is odd.

If n_1 is odd, then the spectral sequence collapses. This result is also contained in Onishchik [**80**]. However, Onishchik uses real cohomology of Lie algebras instead of fibre-bundle techniques and spectral sequences, hence his proof is quite different. The result can also be found in a paper of Hsiang-Su [**47**] about transitive action on Stiefel manifolds. However, Hsiang-Su were obviously not aware of the Cartan-Serre Theorem, because they included a superfluous condition on the rational homotopy groups.

We obtain a relation between degrees of the primitive elements in the cohomology of G and H and the numbers n_1 and n_2. The second important result is that G has a semisimple normal transitive subgroup with at most two almost simple factors.

In the last section we state the classification of all compact 1-connected homogeneous spaces which have the same integral cohomology as a product $\mathbb{S}^{n_1} \times \mathbb{S}^{n_2}$, for $3 \leq n_1 \leq n_2$, n_2 odd. The actual classification is carried out in Chapters 5 and 6, using the results of this chapter and the representation theory developed in the next chapter. This classification is the first main result of this book. However, as I mentioned before, the reader who has a different classification problem in mind should have no problems to adapt the techniques and results of Chapter 3 and 4 for his own purposes.

3.A. Structure of compact Lie groups

We call a compact connected Lie group *almost simple* if it has no normal closed subgroup of positive dimension. Such a group is connected with finite center, and its Lie algebra is simple. The compact simple Lie algebras are the series

$$\begin{aligned} \mathfrak{a}_n &= \mathfrak{su}_{n+1}, \ n \geq 1, \\ \mathfrak{b}_n &= \mathfrak{so}_{2n+1}, \ n \geq 2, \\ \mathfrak{c}_n &= \mathfrak{sp}_n, \ n \geq 3, \\ \mathfrak{d}_n &= \mathfrak{so}_{2n}, \ n \geq 4, \end{aligned}$$

and the five exceptional compact Lie algebras

$$\mathfrak{g}_2, \ \mathfrak{f}_4, \ \mathfrak{e}_6, \ \mathfrak{e}_7, \text{ and } \mathfrak{e}_8.$$

There are isomorphisms

$$\begin{aligned} \mathfrak{su}_2 \cong \mathfrak{so}_3 \cong \mathfrak{sp}_1 \\ \mathfrak{so}_4 \cong \mathfrak{su}_2 \oplus \mathfrak{su}_2 \\ \mathfrak{so}_5 \cong \mathfrak{sp}_2 \\ \mathfrak{so}_6 \cong \mathfrak{su}_4. \end{aligned}$$

The corresponding classical groups are the complex unitary groups

$$\mathrm{SU}(n+1) = \mathrm{SU}_{n+1}\mathbb{C},$$

the orthogonal groups in odd dimensions

$$\mathrm{SO}(2n+1) = \mathrm{SO}_{2n+1}\mathbb{R},$$

the quaternion unitary groups

$$\mathrm{Sp}(n) = \mathrm{U}_n\mathbb{H},$$

and the orthogonal groups in even dimensions

$$\mathrm{SO}(2n) = \mathrm{SO}_{2n}\mathbb{R}.$$

The groups $\mathrm{SU}(n)$ and $\mathrm{Sp}(n)$ are simply connected; the universal coverings of the orthogonal groups $\mathrm{SO}(n)$, $n \geq 3$, are the groups $\mathrm{Spin}(n)$. A compact Lie group of type $\mathfrak{g}_2$, $\mathfrak{f}_4$ or $\mathfrak{e}_8$ is automatically simple and simply connected, and we denote these unique groups by G_2, F_4, and E_8, respectively. The simply connected groups of type $\mathfrak{e}_6, \mathfrak{e}_7$ have centers $\mathbb{Z}/3$ and $\mathbb{Z}/2$, respectively. We denote these groups by E_6 and E_7.

A prefix P denotes the corresponding simple group, e.g. $\mathrm{PSO}(6)$ or PE_6. These groups and the compact 1-torus

$$\mathbb{T} = \mathrm{SO}(2)$$

are the building blocks of all compact connected Lie groups.

THEOREM 3.1 (Structure of compact connected Lie groups).
Let G be a compact connected Lie group. Then there exist simply connected almost simple compact Lie groups $G_1, \dots, G_r$ and a torus $\mathbb{T}^s$ and a surjection

$$f : G_1 \times \cdots \times G_r \times \mathbb{T}^s \longrightarrow G$$

with finite kernel.

PROOF. See e.g. Hofmann-Morris [**44**] Ch. 6, Thm. 6.19, or Salzmann *et al.* [**85**] 93.11. □

A Lie group G is an H-space. Therefore the rational cohomology of G is a finite dimensional associative Hopf algebra,

$$\mathbf{H}^\bullet(G;\mathbb{Q}) \cong \bigwedge\nolimits_{\mathbb{Q}}(u_1,\dots,u_r),$$

cp. Spanier [**90**] 5.8.13, generated by primitive elements $\{u_1,\dots,u_r\}$, i.e. with comultiplication

$$u_i \longmapsto u_i\otimes 1 + 1\otimes u_i, \qquad \text{for } i=1,\dots,r,$$

cp. e.g. Whitehead [**113**] III.8. Since G is an H-space, the fundamental group $\pi_1(G)$ acts trivially on the homotopy groups $\pi_r(G)$, for $r\geq 1$, cp. Whitehead [**113**] III.4.18. To determine the rational cohomology of G, it suffices to consider the case where $s=0$, since G is topologically a product of a $\mathbb{T}^s$ and a semisimple group. So assume that $s=0$. The map f described above induces then an isomorphism

$$\mathbf{H}^\bullet(G_1\times\cdots\times G_r;\mathbb{Q}) \longleftarrow \mathbf{H}^\bullet(G;\mathbb{Q}).$$

(To see the last implication, consider the rational Leray-Serre spectral sequence of the $\mathbb{Q}$-simple fibration

$$G_1\times\cdots\times G_r \longrightarrow G \longrightarrow K(\pi_1,1).$$

The base $K(\pi_1,1)$ is $\mathbb{Q}$-acyclic, since $\pi_1=\pi_1(G)$ is finite — we assumed that $s=0$.) Thus, we see that the Cartan-Serre theorem applies in particular to compact connected Lie groups; in particular, we have an isomorphism

$$\mathbf{H}^\bullet(G_1\times\cdots\times G_r\times\mathbb{T}^s;\mathbb{Q}) \xleftarrow{\cong} \mathbf{H}^\bullet(G;\mathbb{Q})$$

(now s is again arbitrary).

3.2. Put $m_i=\deg(u_i)$. These numbers are known for all compact almost simple Lie groups; they are as follows.

Type	$(m_1,\dots,m_r)$
$\mathfrak{a}_n$	$(3,5,7,\dots,2n+1)$
$\mathfrak{b}_n$	$(3,7,11,\dots,4n-1)$
$\mathfrak{c}_n$	$(3,7,11,\dots,4n-1)$
$\mathfrak{d}_n$	$(3,7,11,\dots,4n-5,2n-1)$
$\mathfrak{e}_6$	$(3,9,11,15,17,23)$
$\mathfrak{e}_7$	$(3,11,15,19,23,27,35)$
$\mathfrak{e}_8$	$(3,15,23,27,35,39,47,59)$
$\mathfrak{f}_4$	$(3,11,15,23)$
$\mathfrak{g}_2$	$(3,11)$

cp. Mimura-Toda [**71**] III.6.5, VI.5.10 or Mimura [**70**] Thm. 2.2. In particular, we see the following: If G is a compact connected Lie group, then the torus factor has dimension $\dim_{\mathbb{Q}}\mathbf{H}^1(G;\mathbb{Q})$, and if G has no torus factors, then the number of almost simple factors is $\dim_{\mathbb{Q}}\mathbf{H}^3(G;\mathbb{Q})$. In fact, it is known that $\pi_3(G)\cong\mathbb{Z}$ for all almost simple compact Lie groups, cp. Mimura [**70**] Thm. 3.9. Let $P_G\subseteq\mathbf{H}^\bullet(G;\mathbb{Q})$ denote the (graded) vector space spanned by the primitive homogeneous generators of the cohomology. We recall the following facts.

- The dimension of P_G is the rank of G.
- The dimension of P_G^1 is the rank of the central torus of G.
- The dimension of P_G^3 is the number of almost simple factors of G.
- If G is almost simple and of rank at least 2, then G is exceptional if and only if $P_G^5=P_G^7=0$ (if the rank of G is at least 3 then it suffices that $P_G^7=0$).

These facts follow essentially from the structure of the rational cohomology of compact almost simple Lie groups given by the table above.

THEOREM 3.3 (Borel). *Let G be a compact connected Lie group, and let $i : H \subseteq G$ be a closed connected subgroup. Then $i^\bullet(P_G)$ is contained in P_H; we denote the restriction of $i^\bullet$ to P_G by*

$$P_i : P_G \longrightarrow P_H.$$

The image $i^\bullet(\mathbf{H}^\bullet(G;\mathbb{Q}))$ is an exterior algebra (and a Hopf algebra) generated by $i^\bullet(P_G)$. The kernel of $i^\bullet$ is the ideal generated by the homogeneous elements contained in it.

PROOF. See Borel [**9**] §21. □

Late, we will need the following result of Borel about compact transformation groups.

THEOREM 3.4 (Borel). *Suppose that $H \subseteq G$ is a closed connected subgroup of a compact Lie group G. Consider the principal bundle*

$$\begin{array}{ccc} H & \longrightarrow & G \\ & & \Big\downarrow p \\ & & G/H \end{array}$$

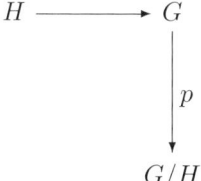

The image $p^\bullet(\mathbf{H}^\bullet(G/H;\mathbb{Q}))$ is an exterior algebra (and a Hopf algebra) generated by $P_G \cap p^\bullet(\mathbf{H}^\bullet(G/H;\mathbb{Q}))$.

PROOF. See Borel [**9**] §21. □

3.B. Certain homogeneous spaces

3.5. IRREDUCIBLE ACTIONS

We begin with some general observations about homogeneous spaces. Let X be a homogeneous space of a compact connected Lie group K. The action of K on X is called *irreducible* if K has no proper normal transitive subgroup. There exists always a normal connected transitive subgroup $G \subseteq K$ such that the action of G on X is irreducible.

Let L be a normal complement of G, i.e. $K = G \cdot L$. Then

$$L \subseteq \mathrm{Cen}_{\mathrm{Sym}(X)}(G) = C$$

is contained in the group C of all permutations of X which centralize G. This group C can be recovered from (G, X) as follows. Let $H = G_x$ be the stabilizer of an element $x \in X$ and put $X = G/H$. We define an action of the normalizer $N = \mathrm{Nor}_G(H)$ on $X = G/H$ by

$$n \cdot (gH) = gn^{-1}H;$$

this action centralizes the action of G on X, and the kernel of this action is H. Thus we have an injection $N/H \hookrightarrow C$. We claim that this injection is an isomorphism

$$C \cong N/H.$$

Let $c \in C$ and note that $H = G_x$ fixes $c(x)$, because C centralizes the action of H. Choose $n \in G$ such that $n^{-1}(x) = c(x)$. For $h \in H$ we have $hn^{-1}(x) = n^{-1}(x)$, so $n \in N$. Now let $y \in X$ be an arbitrary element. Choose $g \in G$ such that $g(x) = y$. Then $c(y) = cg(x) = gc(x) = gn^{-1}(x)$, and this is precisely the N-action on $X = G/H$ described above. Thus $C \cong N/H$. Note that N/H is locally isomorphic to $\mathrm{Cen}_G(H)/\mathrm{Cen}(H)$. Therefore we have the following result.

PROPOSITION 3.6. *Let X be a compact space. To classify all compact connected groups which act transitively and effectively on X, it suffices to classify all pairs (G, H), where G is compact and connected, and where G acts irreducibly on $X = G/H$, and then to determine the G-normalizer (or centralizer) of the isotropy group H.* □

Now we consider a compact 1-connected homogeneous space G/H which has the same integral cohomology as a product of spheres $\mathbb{S}^{n_1} \times \mathbb{S}^{n_2}$, where $n_1 \leq n_2$, and n_2 is odd. Thus, the cohomology of G/H is

$$\mathbf{H}^\bullet(G/H) \cong \bigwedge\nolimits_{\mathbb{Z}}(u, v) \qquad \deg(u) = n_1,\ \deg(v) = n_2$$

if n_1 is odd, and

$$\mathbf{H}^\bullet(G/H) \cong \mathbb{Z}[a]/(a^2) \otimes \bigwedge\nolimits_{\mathbb{Z}}(u) \qquad \deg(a) = n_1,\ \deg(u) = n_2$$

if n_1 is even. We may assume that G is connected, because G/H is connected, and the evaluation map is open.

THEOREM 3.7 (Onishchik). *Let G/H be a 1-connected homogeneous space of a compact connected Lie group G. Suppose that the rational cohomology of G/H is an exterior algebra*

$$\mathbf{H}^\bullet(G/H; \mathbb{Q}) \cong \bigwedge\nolimits_{\mathbb{Q}}(w_1, \ldots, w_r)$$

in homogeneous generators of odd degrees. Let $P_{G/H} \subseteq \mathbf{H}^\bullet(G/H; \mathbb{Q})$ denote the graded vector space spanned by the homogeneous generators $w_1, \ldots, w_r$. Then the spectral sequence of the fibration $G \longrightarrow G/H$ collapses, and there is an exact sequence

$$0 \longleftarrow P_H \longleftarrow P_G \longleftarrow P_{G/H} \longleftarrow 0.$$

In particular,

$$\mathrm{rk}(G) - \mathrm{rk}(H) = r,$$

and we have the following diagram:

$$\begin{array}{ccccc}
\mathbf{H}^\bullet(H;\mathbb{Q}) & \longleftarrow & \mathbf{H}^\bullet(G;\mathbb{Q}) & \longleftarrow & \mathbf{H}^\bullet(G/H;\mathbb{Q}) \\
\downarrow \cong & & \downarrow \cong & & \downarrow \cong \\
\bigwedge_{\mathbb{Q}}(u_1, \ldots, u_s) & \longleftarrow & \bigwedge_{\mathbb{Q}}(u_1, \ldots, u_s, w_1, \ldots, w_r) & \longleftarrow & \bigwedge_{\mathbb{Q}}(w_1, \ldots, w_r).
\end{array}$$

PROOF. Consider the long exact homotopy sequence of the fibration $H \longrightarrow G \longrightarrow G/H$, tensored with $\mathbb{Q}$. All even-dimensional homotopy groups of G, H

and G/H are finite. The field $\mathbb{Q}$ is a flat $\mathbb{Z}$-module, hence there are short exact sequences

$$0 \longrightarrow \pi_k(H) \otimes \mathbb{Q} \longrightarrow \pi_k(G) \otimes \mathbb{Q} \longrightarrow \pi_k(G/H) \otimes \mathbb{Q} \longrightarrow 0$$

for all k (the groups are trivial if k is even). This proves that $\mathrm{rk}(G) = \mathrm{rk}(H) + r$.

The rational spectral sequence of the fibration $H \longrightarrow G \longrightarrow G/H$ has $\mathbf{H}^\bullet(G/H;\mathbb{Q}) \otimes \mathbf{H}^\bullet(H;\mathbb{Q})$ as its $\mathbf{E}_2$-term. Therefore

$$\dim \mathbf{E}_2 = \dim \mathbf{H}^\bullet(G/H;\mathbb{Q}) \cdot \dim \mathbf{H}^\bullet(H;\mathbb{Q}) = 2^r \cdot 2^{\mathrm{rk}(H)}.$$

But

$$\dim \mathbf{E}_\infty = \dim \mathbf{H}^\bullet(G;\mathbb{Q}) = 2^{\mathrm{rk}(G)} = 2^{r+\mathrm{rk}(H)} = \dim \mathbf{E}_2.$$

Thus, the spectral sequence collapses. This implies that the map $H \longrightarrow G$ induces a surjection in cohomology, and that $G \longrightarrow G/H$ induces an injection. The claim follows from 3.3 and 3.4 and the Leray-Hirsch Theorem 1.17. □

The result above follows also from Theorem 1 on p. 216 in Onishchik [**80**] (for cohomology with real coefficients).

3.8. How to use this

Let G/H be a homogeneous space as in the theorem above. Let

$$(m_1^G, \dots, m_{r+s}^G)$$

denote the degrees of the homogeneous generators of the rational cohomology of G, and let

$$(m_1^H, \dots, m_s^H)$$

denote the corresponding numbers for H. Then, after a suitable permutation of the indices, $m_i^G = m_i^H$ for $i = 1, \dots, s$, so $(m_1^H, \dots, m_s^H)$ has to be a subsequence of $(m_1^G, \dots, m_{r+s}^G)$, and the sequence which remains after deleting all entries of the subsequence gives the degrees of the homogeneous generators of the cohomology of G/H. This is essentially the method that we are going to use in the case that G is an almost simple group. For example, if $G = \mathrm{SU}(5)$ and $H = \mathrm{SU}(3)$, then we obtain

$$\begin{array}{rl} m^G: & (\not{3}, \not{5}, 7, 9) \\ m^H: & (3, 5) \\ \hline & (7, 9) \end{array}$$

If G is not almost simple, we need an additional reduction method, i.e. a bound on the number of almost simple factors (see 3.14 below).

Now we assume that the cohomology of $X = G/H$ is

$$\mathbf{H}^\bullet(X;\mathbb{Q}) \cong \mathbb{Q}[a]/(a^2) \otimes \bigwedge(u)$$

where $\deg(a) = n_1$ is even, $\deg(u) = n_2$ is odd, and $n_1 < n_2$. There are short exact sequences

$$0 \longrightarrow \pi_k(H) \otimes \mathbb{Q} \longrightarrow \pi_k(G) \otimes \mathbb{Q} \longrightarrow \pi_k(G/H) \otimes \mathbb{Q} \longrightarrow 0$$

for $k \neq n_1 - 1$, and an exact sequence

$$0 \longrightarrow \pi_{n_1}(G/H) \otimes \mathbb{Q} \longrightarrow \pi_{n_1-1}(H) \otimes \mathbb{Q} \longrightarrow \pi_{n_1-1}(G) \otimes \mathbb{Q} \longrightarrow 0,$$

since $\pi_{n_1-1}(G/H)$ is finite. This shows that $\mathrm{rk}(G) - \mathrm{rk}(H) = 1$. The rational Leray-Serre spectral sequence of the fibration $H \longrightarrow G \longrightarrow G/H$ is slightly more complicated because there are non-trivial differentials. We have

$$\dim \mathbf{E}_2 = \dim \mathbf{H}^\bullet(G/H;\mathbb{Q}) \cdot \dim \mathbf{H}^\bullet(H;\mathbb{Q}) = 4 \cdot 2^{\mathrm{rk}(H)},$$

and

$$\dim \mathbf{E}_\infty = \dim \mathbf{H}(G;\mathbb{Q}) = 2^{\mathrm{rk}(G)} = 2 \cdot 2^{\mathrm{rk}(H)}.$$

This shows already that the spectral sequence does not collapse. The terms $\mathbf{E}_2^{s,t}$ in the strip $0 < s < n_1$ are zero, because $\mathbf{H}^s(G/H;\mathbb{Q}) = 0$ for these values of s. Therefore d_k is trivial on $\mathbf{E}_k^{0,\bullet}$ for $k < n_1$ and this implies by 1.15 that d_k is trivial for $k = 2, \dots, n_1 - 1$ on all terms of $\mathbf{E}_k$. Recall that the following diagram commutes.

$$\begin{array}{ccccccc}
\mathbf{H}^\bullet(G/H;\mathbb{Q}) & & \longrightarrow & & & & \mathbf{H}^\bullet(G;\mathbb{Q}) \\
\downarrow \cong & & & & & & \| \\
\mathbf{H}^\bullet(G/H;\mathbb{Q}) \otimes 1 & = & \mathbf{E}_2^{\bullet,0} & \longrightarrow & \mathbf{E}_\infty^{\bullet,0} & \hookrightarrow & \mathbf{H}^\bullet(G;\mathbb{Q})
\end{array}$$

Now we use the fact that the image of the cohomology of G/H is generated by the primitive elements contained in it. A primitive element has odd degree. The possibly non-zero terms of odd degree in $\mathbf{E}_\infty^{\bullet,0}$ are $\mathbf{H}^{n_2}(G/H;\mathbb{Q})$ and $\mathbf{H}^{n_1+n_2}(G/H;\mathbb{Q})$. They cannot generate an element of degree n_1, hence $\mathbf{E}_\infty^{n_1,0} = 0$. In the spectral sequence, the only possibly non-zero arrow which ends at $\mathbf{E}_k^{n_1,0}$ is

$$d_{n_1} : \mathbf{E}_{n_1}^{0,n_1-1} \longrightarrow \mathbf{E}_{n_1}^{n_1,0}.$$

Thus this map is surjective. Pick an element $w_1 \in \mathbf{H}^{n_1-1}(H;\mathbb{Q})$ with $d_{n_1}(1 \otimes w_1) = a \otimes 1$. All primitive elements in $\mathbf{H}^\bullet(H;\mathbb{Q})$ of degree less than $n_1 - 1$ are mapped to zero. Therefore, w_1 is not a sum of products of primitive elements of lower degree, i.e. w_1 itself is a primitive element. Put

$$\mathbf{H}^\bullet(H;\mathbb{Q}) = \textstyle\bigwedge_{\mathbb{Q}}(w_1, w_2, \dots w_r),$$

where $\mathrm{rk}(H) = r$. We claim that the elements $w_2, \dots, w_r$ can be chosen in such a way that the following holds.

(1) $d_{n_1}(1 \otimes w_i) = 0$ for $i = 2, \dots, r$
(2) The spectral sequence collapses at $n_1 + 1$, i.e. $\mathbf{E}_{n_1+1} = \mathbf{E}_\infty$.

For the second claim, it suffices to show that $\dim \mathbf{E}_{n_1+1} \leq 2^{r+1}$, since we know already that $\dim \mathbf{E}_{n_1+1} \geq \dim \mathbf{E}_\infty = 2^{r+1}$, and the equality $\dim \mathbf{E}_{n_1+1} = \dim \mathbf{E}_\infty$ forces that $d_k = 0$ for all $k \geq n_1 + 1$. Since $\mathbf{E}_{n_1+1}$ is the cohomology of $\mathbf{E}_{n_1}$, we have

$$\dim \mathbf{E}_{n_1+1} = \dim \mathbf{E}_{n_1} - 2\dim(\mathrm{im}(d_{n_1})).$$

So we need to show that $\dim(\mathrm{im}(d_{n_1})) \geq 2^r$ because then $\dim \mathbf{E}_{n_1+1} \leq 4\cdot 2^r - 2\cdot 2^r = 2^{r+1}$. Define a linear map ϕ by $d_{n_1}(1 \otimes z) = a \otimes \phi(z)$. Then

$$\begin{aligned}
d_{n_1}(1 \otimes w_1 z) &= d_{n_1}(1 \otimes w_1)(1 \otimes z) - (1 \otimes w_1) d_{n_1}(1 \otimes z) \\
&= (a \otimes 1)(1 \otimes z) - (1 \otimes w_1)(a \otimes \phi(z)) \\
&= a \otimes (z - w_1 \phi(z)).
\end{aligned}$$

Now $w_1(z - w_1\phi(z)) = w_1 z \neq 0$, provided that $z \in \bigwedge_{\mathbb{Q}}(w_2, \dots, w_r)$. Thus, d_{n_1} is injective on the 2^{r-1}-dimensional space $1 \otimes \left(w_1 \bigwedge_{\mathbb{Q}}(w_2, \dots, w_r)\right)$. For $u \otimes z \in \mathbf{E}_2^{n_2,\bullet}$ we have similarly

$$\begin{aligned} d_{n_1}(u \otimes z) &= d_{n_1}(u \otimes 1)(1 \otimes z) - (u \otimes 1)d_{n_1}(1 \otimes z) \\ &= -(u \otimes 1)(a \otimes \phi(z)) \\ &= -(ua) \otimes \phi(z). \end{aligned}$$

Therefore, d_{n_1} is also injective on the 2^{r-1}-dimensional subspace

$$u \otimes \left(w_1 \bigwedge\nolimits_{\mathbb{Q}}(w_2, \dots, w_r)\right).$$

The images of $1 \otimes \left(w_1 \bigwedge_{\mathbb{Q}}(w_2, \dots, w_r)\right)$ and $u \otimes \left(w_1 \bigwedge_{\mathbb{Q}}(w_2, \dots, w_r)\right)$ have trivial intersection (because of the grading), therefore the image of d_{n_1} has dimension

$$\dim(\mathrm{im}(d_{n_1})) = 2^r.$$

We have proved claim (2). Moreover, we know the cohomology of the map $G \xrightarrow{p} G/H$, and we use this to prove claim (1).

LEMMA 3.9. *The map* $\mathbf{H}^\bullet(G;\mathbb{Q}) \longleftarrow \mathbf{H}^\bullet(G/H;\mathbb{Q})$ *is given by the diagram*

$$\begin{array}{ccc} \mathbf{H}^\bullet(G;\mathbb{Q}) & \xleftarrow{\quad p^\bullet \quad} & \mathbf{H}^\bullet(G/H;\mathbb{Q}) \\ \| & & \| \\ \bigwedge_{\mathbb{Q}}(u, v_1, \dots, v_r) & \longleftarrow & \mathbb{Q}[a]/(a^2) \otimes \bigwedge_{\mathbb{Q}}(u), \end{array}$$

where $p^\bullet(u) = u$ *and* $p^\bullet(a) = 0$. □

We still have not yet determined the differential d_{n_1}. We apply 3.3 to the inclusion $h : H \subseteq G$. Recall that the following diagram commutes.

$$\begin{array}{ccccc} \mathbf{H}^\bullet(G;\mathbb{Q}) & & \xrightarrow{\quad h^\bullet \quad} & & \mathbf{H}^\bullet(H;\mathbb{Q}) \\ \downarrow & & & & \| \\ \mathbf{E}_\infty^{0,\bullet} & \hookrightarrow & \mathbf{E}_2^{0,\bullet} & \xrightarrow{\cong} & \mathbf{H}^\bullet(H;\mathbb{Q}) \end{array}$$

Thus, we can identify the image of $h^\bullet$ with $\mathbf{E}_\infty^{0,\bullet} \subseteq \mathbf{E}_2^{0,\bullet}$. We know that $\dim \mathbf{E}_\infty^{0,\bullet} = 2^{r-1}$. Therefore $\dim(\mathrm{im}(h^\bullet) \cap P_H) = r - 1$, and we can find primitive elements $w_2, \dots w_r \in P_H$ which generate $\mathrm{im}(h^\bullet) \cong \mathbf{E}_\infty^{0,\bullet}$. This implies that $d_{n_1}(1 \otimes w_i) = 0$ for $i = 2, \dots, r$.

LEMMA 3.10. *There exist primitive elements* $w_1, \dots w_r$ *which generate* $\mathbf{H}^\bullet(H;\mathbb{Q})$ *such that*

$$d_{n_1}(1 \otimes w_1) = a \otimes 1, \qquad \text{and} \qquad d_{n_1}(1 \otimes w_i) = 0 \text{ for } i = 2, \dots, r.$$

□

Consider the map $P_G \longrightarrow P_H$. Its image has dimension $(r-1)$, thus its kernel has dimension 2. We may choose an $(r-1)$-dimensional subspace of P_G which maps isomorphically onto $h^\bullet(P_G)$. The image of u is contained in $\mathbf{E}_\infty^{n_1,0}$ and thus in the kernel of $h^\bullet$. We choose one more primitive element v such that u, v span the kernel of $P_G \longrightarrow P_H$. Note that $\deg(v) = 2n_1 - 1$ by 2.4.

THEOREM 3.11. *Let $X = G/H$ be a 1-connected homogeneous space of a compact connected Lie group G. Suppose that the rational cohomology of G/H is of the form*

$$\mathbf{H}^\bullet(G/H;\mathbb{Q}) \cong \mathbb{Q}[a]/(a^2) \otimes \bigwedge\nolimits_{\mathbb{Q}}(u),$$

where $\deg(a)$ *is even and* $\deg(u)$ *is odd, and* $\deg(a) < \deg(u)$. *Then* $\mathrm{rk}(G) - \mathrm{rk}(H) = 1$, *and the following diagram gives the induced maps in cohomology,*

$$\begin{array}{ccccc} \mathbf{H}^\bullet(H;\mathbb{Q}) & \xleftarrow{\ h^\bullet\ } & \mathbf{H}^\bullet(G;\mathbb{Q}) & \xleftarrow{\ p^\bullet\ } & \mathbf{H}^\bullet(G/H;\mathbb{Q}) \\ \Big\| & & \Big\| & & \Big\| \\ \bigwedge_{\mathbb{Q}}(w_1,\dots,w_r) & \longleftarrow & \bigwedge_{\mathbb{Q}}(w_2,\dots,w_r,u,v) & \longleftarrow & \mathbb{Q}[a]/(a^2)\otimes\bigwedge_{\mathbb{Q}}(u) \end{array}$$

where $p^\bullet(a) = 0$, and $h^\bullet(u) = 0 = h^\bullet(v)$. Moreover, $\deg(v) = 2\deg(a) - 1$.

3.12. HOW TO USE THIS

Let G/H be a homogeneous space as in the theorem above. Let

$$(m_1^G, \dots, m_{s+1}^G)$$

denote the degrees of the homogeneous generators of the rational cohomology of G, and let

$$(m_1^H, \dots, m_s^H)$$

denote the corresponding numbers for H. Then, after a suitable permutation of the indices, $m_i^G = m_i^H$ for $i = 1, \dots, s-1$, and $m_s^G = 2m_s^H + 1$ so $(m_1^H, \dots, m_{s-1}^H, 2m_s^H + 1)$ has to be a subsequence of $(m_1^G, \dots, m_{s+1}^G)$. The homogeneous generators of the cohomology of G/H then have degrees $n_1 = m_s^H + 1$ and $n_2 = m_{s+1}^G$. Here is an example: $G = \mathrm{SU}(5)$ and $H = \mathrm{SU}(3) \times \mathrm{SU}(2)$.

$$\begin{array}{llcl} m^G : & & & (\not{3}, \not{5}, \not{7}, 9) \\ m^H : & (3,5,3) & \rightsquigarrow & (3,5,\boxed{7}) \\ \hline & & & (\boxed{4}, 9) \end{array}$$

We want to show that there exists a transitive normal semisimple subgroup of G with at most 2 almost simple factors. To prove this, we use the following result.

LEMMA 3.13 (Hsiang-Su, Onishchik). *Let G/H be a homogeneous space of a compact connected Lie group G, and let $N \subseteq G$ be a closed connected normal subgroup. Let $\eta : H \subseteq G$ and $\nu : N \subseteq G$ be the inclusion maps. Let P_η and P_ν denote the induced maps on the vector spaces of primitive elements. If* $\ker(P_\eta)$ *injects into P_N under P_ν, then N acts transitively on X.*

PROOF. Passing to a suitable compact covering, we may assume that $G \cong N \times G/N$. The maps $N \longrightarrow G \longrightarrow G/N$ yield a short exact sequence

$$0 \longleftarrow P_N \longleftarrow P_G \longleftarrow P_{G/N} \longleftarrow 0,$$

and we obtain a diagram

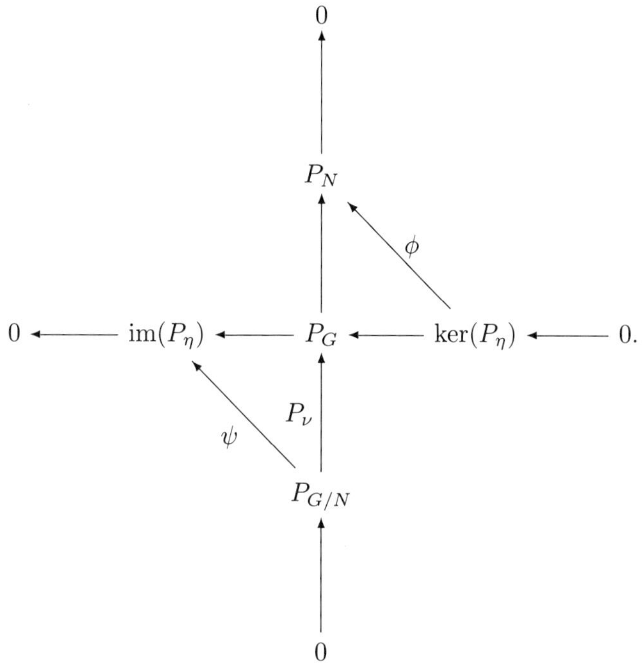

The row and the column are exact, hence ϕ is injective if and only if ψ is injective. By our assumption ϕ is injective, hence the composite $H \hookrightarrow G \longrightarrow G/N$ induces an injection

$$P_H \longleftarrow P_{G/N}.$$

Therefore, the map $\mathbf{H}^\bullet(H;\mathbb{Q}) \longleftarrow \mathbf{H}^\bullet(G/N;\mathbb{Q})$ is also an injection. We factor it through the image $HN/N \cong H/H \cap N$ of H in G/N to obtain

$$\mathbf{H}^\bullet(H;\mathbb{Q}) \longleftarrow \mathbf{H}^\bullet(HN/N;\mathbb{Q}) \longleftarrow \mathbf{H}^\bullet(G/N;\mathbb{Q}).$$

In particular, the map on the right must be an injection. But $HN/N \subseteq G/N$; therefore, the map on the right is an isomorphism, and $HN/N = G/N$. Thus $HN = G$, and N acts transitively on G/H. □

A similar result is proved in Hsiang-Su [**47**] Prop.1.4 and in Onishchik [**80**] §17.1.

PROPOSITION 3.14. *Let G/H be a homogeneous space whose cohomology is either of the form*

$$\bigwedge\nolimits_{\mathbb{Z}}(u, v),$$

where $\deg(u), \deg(v) \geq 3$ *are odd, or of the form*

$$\mathbb{Z}[a]/(a^2) \otimes \bigwedge\nolimits_{\mathbb{Z}}(u).$$

where $\deg(a)$ *is even and* $\deg(u)$ *is odd, and* $\deg(u) > \deg(a) \geq 4$. *There exists a normal transitive semisimple subgroup* $N \subseteq G$ *with at most two almost simple factors.*

PROOF. Passing to a suitable covering, we may assume that G is a product of torus groups and simply connected almost simple groups,

$$G = G_1 \times \cdots \times G_s.$$

Then $P_G = P_{G_1} \oplus \cdots \oplus P_{G_s}$. The kernel of $P_\eta : P_G \longrightarrow P_H$ is 2-dimensional. The inclusion $G_i \hookrightarrow G$ induces the projection $P_G \longrightarrow P_{G_i}$. There exist numbers i, j such that $\ker(P_\eta)$ injects into $P_{G_i} \oplus P_{G_j}$ under the map induced from $G_i \times G_j \hookrightarrow G$. Therefore, $N = G_i \times G_j$ acts transitively on G/H by 3.13. □

3.C. The integral classification

We use the following terminology. The point space of the n-dimensional projective geometry over the skew field $\mathbb{F}$ is denoted by

$$\mathbb{F}\mathrm{P}^n,$$

for $\mathbb{F} = \mathbb{R}, \mathbb{C}, \mathbb{H}$. The point space of the projective Cayley plane is denoted by $\mathbb{O}\mathrm{P}^2$. The *Stiefel manifold* of orthonormal k-frames in $\mathbb{F}^n$ is

$$V_k(\mathbb{F}^n) = \{(v, \dots, v_k) \in \mathbb{F}^{kn} | \ (v_i|v_j) = \delta_{i,j}\},$$

where $(x|y) = \sum_{i=1}^n \bar{x}_i y_i$, for $\mathbb{F} = \mathbb{R}, \mathbb{C}, \mathbb{H}, \mathbb{O}$ (for $\mathbb{F} = \mathbb{O}$ we restrict k to $k = 1, 2$ since $V_k(\mathbb{O}^n)$ is a singular algebraic variety and not a manifold for $k > 2$). We define the *oriented Grassmann manifolds* as

$$\widetilde{G}_k(\mathbb{R}^n) = \mathrm{SO}(n)/\mathrm{SO}(k) \times \mathrm{SO}(n-k) \text{ and } \widetilde{G}_k(\mathbb{C}^n) = \mathrm{SU}(n)/\mathrm{SU}(k) \times \mathrm{SU}(n-k).$$

In Chapter 5 and 6 we prove the following main result.

THEOREM 3.15. *Let* $X = G/H$ *be a 1-connected homogeneous space of a compact connected Lie group* G. *Suppose that the action of* G *on* X *is irreducible, cp. 3.5.*

(A) *Assume that* X *has the same rational cohomology as a product of spheres*

$$\mathbb{S}^{n_1} \times \mathbb{S}^{n_2},$$

where $n_2 \geq n_1 \geq 3$ *and* n_2 *is odd. Then* (G, H) *is one of the pairs which we discuss in chapters 5-6.*

(B) *Suppose that* X *has the same integral cohomology as a product of spheres*

$$\mathbb{S}^{n_1} \times \mathbb{S}^{n_2},$$

where $n_2 \geq n_1 \geq 3$ *and* n_2 *is odd. There are the following possibilities (and no others).*

(B1) *If* n_1 *is odd, then* (G, H) *is one of the pairs in 5.6, 6.15, or 6.7. More precisely we have the following spaces.*

Stiefel manifolds

$$\mathrm{SU}(n)/\mathrm{SU}(n-2) = V_2(\mathbb{C}^n), \quad n \geq 3$$
$$\mathrm{Sp}(n)/\mathrm{Sp}(n-2) = V_2(\mathbb{H}^n), \quad n \geq 2$$

Homogeneous sphere bundles

$$\mathrm{Sp}(n) \times \mathrm{SU}(3)/\mathrm{Sp}(n-1) \cdot \mathrm{Sp}(1)$$
$$\mathrm{Sp}(n) \times \mathrm{Sp}(2)/\mathrm{Sp}(n-1) \cdot \mathrm{Sp}(1)$$

Products of homogeneous spheres

$$K_1/H_1 \times K_2/H_2$$

where K_1/H_1 and K_2/H_2 is one of the spaces

$$\mathrm{SO}(2n)/\mathrm{SO}(2n-1) = \mathbb{S}^{2n-1}$$
$$\mathrm{SU}(n)/\mathrm{SU}(n-1) = \mathbb{S}^{2n-1}$$
$$\mathrm{Sp}(n)/\mathrm{Sp}(n-1) = \mathbb{S}^{4n-1}$$
$$\mathrm{Spin}(9)/\mathrm{Spin}(7) = \mathbb{S}^{15}$$
$$\mathrm{Spin}(7)/\mathrm{G}_2 = \mathbb{S}^{7}$$

Sporadic spaces

$$\mathrm{E}_6/\mathrm{F}_4$$
$$\mathrm{Spin}(10)/\mathrm{Spin}(7)$$
$$\mathrm{Spin}(9)/\mathrm{G}_2 = V_2(\mathbb{O}^2)$$
$$\mathrm{Spin}(8)/\mathrm{G}_2 = \mathbb{S}^7 \times \mathbb{S}^7$$
$$\mathrm{SU}(6)/\mathrm{Sp}(3) = \mathrm{SU}(5)/\mathrm{Sp}(2)$$

(B2) *If $n_1 \geq 4$ is even, then (G, H) is one of the pairs in 5.14, 6.20, or 6.7. Thus, G/H is one of the following spaces.*

Stiefel manifolds

$$\mathrm{SO}(2n)/\mathrm{SO}(2n-2) = V_2(\mathbb{R}^{2n}), \quad n \geq 3$$

Homogeneous sphere bundles

$$\mathrm{Sp}(n) \times \mathrm{Sp}(2)/\mathrm{Sp}(n-1) \cdot \mathrm{Sp}(1) \cdot \mathrm{Sp}(1)$$

Products of homogeneous spheres

$$K_1/H_1 \times K_2/H_2$$

where K_1/H_1 is one of the spaces

$$\mathrm{SO}(2n+1)/\mathrm{SO}(2n) = \mathbb{S}^{2n}$$
$$\mathrm{G}_2/\mathrm{SU}(3) = \mathbb{S}^{6}$$

and K_2/H_2 is one of the spaces

$$\mathrm{SO}(2n)/\mathrm{SO}(2n-1) = \mathbb{S}^{2n-1}$$
$$\mathrm{SU}(n)/\mathrm{SU}(n-1) = \mathbb{S}^{2n-1}$$
$$\mathrm{Sp}(n)/\mathrm{Sp}(n-1) = \mathbb{S}^{4n-1}$$
$$\mathrm{Spin}(9)/\mathrm{Spin}(7) = \mathbb{S}^{15}$$
$$\mathrm{Spin}(7)/\mathrm{G}_2 = \mathbb{S}^{7}$$

Sporadic spaces

$$\mathrm{Spin}(10)/\mathrm{SU}(5) = \mathrm{Spin}(9)/\mathrm{SU}(4)$$
$$\mathrm{Spin}(7)/\mathrm{SU}(3) = V_2(\mathbb{R}^8)$$
$$\mathrm{Sp}(3)/\mathrm{Sp}(1) \times \mathrm{Sp}(1)$$
$$\mathrm{Sp}(3)/\mathrm{Sp}(1) \times {}^{\mathbb{H}}\rho_{3\lambda_1}(\mathrm{Sp}(1))$$
$$\mathrm{SU}(5)/\mathrm{SU}(3) \times \mathrm{SU}(2)$$

(see Chapter 4 for the definition of ${}^{\mathbb{H}}\rho_{3\lambda_1} : \mathrm{Sp}(1) \longrightarrow \mathrm{Sp}(3)$*).*

(C) *Suppose that G/H is an $(m-1)$-connected rational cohomology $(2m+1)$-sphere, with $\pi_{m-1}(G/H) \cong \mathbb{Z}/2$, and that $m \geq 3$. Then $m = 2n-1$ is odd, $G/H = V_2(\mathbb{R}^{2n+1})$ is a real Stiefel manifold, and G/H is one of the pairs*

$$\mathrm{SO}(2n+1)/\mathrm{SO}(2n-1) = V_2(\mathbb{R}^{2n+1}), \quad n \geq 2$$
$$\mathrm{G}_2/\mathrm{SU}(2) = V_2(\mathbb{R}^7),$$

cp. 6.8. □

It is interesting to picture the distribution of the possible values for (n_1, n_2). Of course, one has to disregard the products of homogeneous spheres, since for them every pair (n_1, n_2) is possible. The diagram below displays the numbers $(n_1, n_2 - n_1)$. The three horizontal infinite series are the Stiefel manifolds and the three vertical series are the homogeneous sphere bundles; they are marked as white circles. The sporadic cases are marked as black dots. A double circle ⊚ or double dot indicates that there are two irreducible group actions which yield the same numbers $(n_1, n_2 - n_1)$.

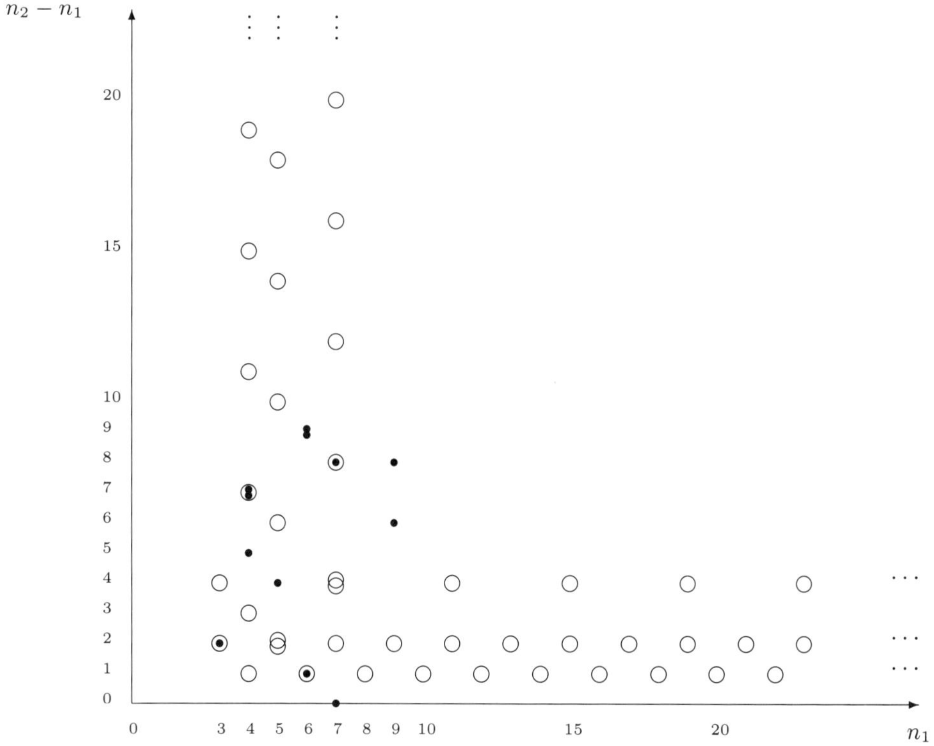

An immediate consequence is the following result.

COROLLARY 3.16. *Let $X = G/H$ be a homogeneous space as in Theorem 3.15 (B). Assume that X is 9-connected, or equivalently, that $n_1 > 9$. Then X is either a product of homogeneous spheres, or X is a Stiefel manifold.* □

Onishchik [**80**] defines the *rank* of a homogeneous space G/H as

$$\mathrm{rk}(G/H) = \dim(\ker[P_H \longleftarrow P_G]).$$

The homogeneous spaces G/H classified in Chapter 5 and 6 are therefore homogeneous of rank 1 or 2. Onishchik's book [**80**] contains tables (on p. 265 and p. 270) of homogeneous spaces of rank 1 (on p. 265) and rank 2 (on p. 270), provided that G is almost simple (and for the case of rank 2, if H is also simple). Our classification shows that the table on p. 270 *loc. cit.* is incomplete; the spaces Spin(11)/Spin(7) and Spin(10)/Spin(7) are missing, cp. 5.2 and 5.4.

CHAPTER 4

Representations of compact Lie groups

In this chapter we collect the necessary facts about (finite dimensional) representations of compact semisimple Lie groups. First we describe the irreducible representations on real, complex and quaternionic vector spaces, and we explain how the problem of classifying certain subgroups of compact classical groups can be reduced to representation theory.

The next section gives tables of the fundamental representations of all compact almost simple Lie groups, and also of all low-dimensional irreducible representations. The last section introduces an algebraic invariant, the Dynkin index, of a homomorphism between almost simple compact Lie groups.

I believe that these results and tables are of some independent interest for geometers.

The representation theory is based on Tits [**105**], [**104**], and on the appendices in Onishchik-Vinberg [**81**]. The section about the Dynkin index is taken from Onishchik [**80**] p. 58–61.

Throughout this chapter G is a compact semisimple Lie group. Such a group may be viewed as an algebraic semisimple $\mathbb{R}$-group, and all results about rational representations of algebraic groups apply.

4.A. The classification of irreducible representations

The material in this section is taken from Tits [**104**] and Tits [**105**], see in particular [**105**] 7.2 and 8.1.1.

Let D, E be central finite dimensional (skew) fields over $\mathbb{R}$, i.e. $D, E \in \{\mathbb{R}, \mathbb{C}, \mathbb{H}\}$, with $D \subseteq E$. Let X be a finite dimensional E-module. The ring of all E-linear endomorphisms of X is denoted by

$$\mathrm{End}_E(X).$$

The group of invertible endomorphisms is denoted by $\mathrm{GL}_E(X) = \mathrm{End}_E(X)^\times$.

We may also view X as a D-module by *restriction of scalars*; this module is denoted by

$$\mathrm{rest}_D^E(X),$$

and we put $\mathrm{End}_D(X) = \mathrm{End}_D(\mathrm{rest}_D^E(X))$ for short. Clearly, $\mathrm{End}_E(X) \subseteq \mathrm{End}_D(X)$. Conversely, if Y is a D-module, then we can consider the E-module

$$\mathrm{tens}_E^D(Y) = Y \otimes_D E$$

obtained by *extension of scalars*.

A continuous group homomorphism $\rho : G \longrightarrow \mathrm{GL}_E(X)$ is called an E-representation of G on X; we call X a G-E-module. A homomorphism between two

G-E-modules is a E-linear map which is G-equivariant. In particular, we have the ring

$$\mathrm{End}_E^G(X) = \mathrm{Cen}_{\mathrm{End}_E(X)}(\rho(G))$$

of all G-equivariant endomorphisms of X. We can apply the above functors rest and tens to G-E-modules; then $\mathrm{rest}_D^E(X)$ is a G-D-module, and if Y is a G-D-module, then $\mathrm{tens}_E^D(Y)$ is a G-E-module.

A non-zero G-E-module is called (E-)*simple* if it has no proper non-zero G-E-submodule, and *semisimple* if it decomposes into a direct sum of simple G-E-modules. Since G is compact, every G-E-module is semisimple.

If two G-E-modules are isomorphic, then the corresponding representations are called *equivalent. Note that equivalence depends on the ground field E.* Let G, G' be two compact semisimple Lie groups, with E-modules X, X'. If there exists an isomorphism $f : G \longrightarrow G'$ and an E-linear isomorphism $F : X \longrightarrow X'$ such that $f(g)F(x) = F(gx)$, then the modules are called *quasi-isomorphic*, and the representations of G and G' are called *quasi-equivalent.*

We recall the following facts from complex representation theory. Let Λ denote the *weight lattice* of G, and let Λ_+ denote the set of *dominant weights.* The weight lattice is generated by *fundamental weights* $\lambda_1, \ldots, \lambda_k$, where k is the rank of G,

$$\Lambda = \bigoplus_{i=1}^{k} \lambda_i \mathbb{Z}.$$

The dominant weights are the weights with non-negative coefficients,

$$\Lambda_+ = \bigoplus_{i=1}^{k} \lambda_i \mathbb{N}_0.$$

4.1. Classification of simple G-$\mathbb{C}$-modules

To each dominant weight $\lambda \in \Lambda_+$ one can associate a $\mathbb{C}$-module X_λ and an irreducible representation $\rho_\lambda : G \longrightarrow \mathrm{End}_\mathbb{C}(X_\lambda)$. Conversely, every simple G-$\mathbb{C}$-module is $\mathbb{C}$-isomorphic to a unique module of this type. Thus, Λ_+ corresponds bijectively to the isomorphism classes of irreducible G-$\mathbb{C}$-modules. The representations ρ_{λ_i} corresponding to the fundamental weights are called the *fundamental representations.*

To classify the real and quaternionic G-modules one needs some additional structure.

4.2. The action of the Galois group

The Galois group $\Gamma = \mathrm{Gal}(\mathbb{C}/\mathbb{R})$ acts on the collection of all G-$\mathbb{C}$-modules as follows. If $\rho : G \longrightarrow \mathrm{End}_\mathbb{C}(X)$ is a representation, then $\bar{\rho} : G \longrightarrow \mathrm{End}_\mathbb{C}(X)$ is given by applying complex conjugation to each entry of the matrix $\rho(g)$ (with respect to some basis of X). There is a corresponding action of Γ on the semigroup Λ_+, i.e. $\overline{\rho_\lambda} = \rho_{\bar{\lambda}}$. This action is additive, hence it suffices to know $\bar{\lambda}_i$ for $i = 1, \ldots, k$.

Let $\lambda \in \Lambda_+$, and let $X = X_\lambda$ be the corresponding simple G-$\mathbb{C}$-module. A *real structure* on X is a conjugate-linear involution $\theta \in \mathrm{End}_\mathbb{R}^G(X)$, i.e.

$$\theta^2 = 1 \neq \theta \quad \text{and} \quad \theta(xa) = \theta(x)\bar{a}$$

for all $x \in X$ and $a \in \mathbb{C}$. If such a real structure exists, then ${}^{\mathbb{R}}X = \mathrm{Fix}(\theta, X_\lambda)$ is a real G-invariant subspace. Moreover, $X = \mathrm{tens}_{\mathbb{C}}^{\mathbb{R}}({}^{\mathbb{R}}X)$. We denote the representation of G on ${}^{\mathbb{R}}X$ by

$$ {}^{\mathbb{R}}\rho_\lambda : G \longrightarrow \mathrm{End}_{\mathbb{R}}({}^{\mathbb{R}}X). $$

If no such real structure exists on X, then we put ${}^{\mathbb{R}}X = \mathrm{rest}_{\mathbb{R}}^{\mathbb{C}}(X)$ and

$$ {}^{\mathbb{R}}\rho_\lambda = \mathrm{rest}_{\mathbb{R}}^{\mathbb{C}}(\rho_\lambda) : G \longrightarrow \mathrm{End}_{\mathbb{R}}({}^{\mathbb{R}}X). $$

4.3. Classification of simple G-$\mathbb{R}$-modules

Every simple G-$\mathbb{R}$-module Y is isomorphic to a module ${}^{\mathbb{R}}X$ obtained by the method described above. Moreover, if ${}^{\mathbb{R}}\rho$ and ${}^{\mathbb{R}}\sigma$ are $\mathbb{R}$-equivalent irreducible representations, then $\rho = \sigma$ or $\rho = \bar{\sigma}$. Thus, the $\mathbb{R}$-irreducible G-$\mathbb{R}$-modules correspond bijectively to the orbit space Λ_+/Γ.

A *quaternionic structure* on a G-$\mathbb{C}$-module X is a semilinear map $\theta \in \mathrm{End}_{\mathbb{R}}^{G}(X)$ with

$$ \theta^2 = -1 \quad \text{and} \quad \theta(xa) = \theta(x)\bar{a} $$

for all $x \in X$ and $a \in \mathbb{C}$. If such a quaternionic structure exists, then the ring generated over $\mathbb{R}$ by θ and $x \longmapsto xi$ is clearly isomorphic to $\mathbb{H}$. We denote the corresponding G-$\mathbb{H}$-module by ${}^{\mathbb{H}}X$. Thus, $X = \mathrm{rest}_{\mathbb{C}}^{\mathbb{H}}({}^{\mathbb{H}}X)$, and

$$ {}^{\mathbb{H}}\rho : G \longrightarrow \mathrm{End}_{\mathbb{H}}({}^{\mathbb{H}}X) $$

is defined by $\rho = \mathrm{rest}_{\mathbb{C}}^{\mathbb{H}}({}^{\mathbb{H}}\rho)$. If no such quaternionic structure exists, then we put ${}^{\mathbb{H}}X = \mathrm{tens}_{\mathbb{H}}^{\mathbb{C}}(X)$ and ${}^{\mathbb{H}}\rho = \mathrm{tens}_{\mathbb{H}}^{\mathbb{C}}(\rho)$.

4.4. Classification of simple G-$\mathbb{H}$-modules

Every simple G-$\mathbb{H}$-module W is isomorphic to a module ${}^{\mathbb{H}}Y$ obtained by the method described above. Moreover, if ${}^{\mathbb{H}}\rho$ and ${}^{\mathbb{H}}\sigma$ are $\mathbb{H}$-equivalent irreducible representations, then $\rho = \sigma$ or $\rho = \bar{\sigma}$. Thus, the $\mathbb{H}$-irreducible G-$\mathbb{H}$-modules correspond bijectively to the orbit space Λ_+/Γ.

4.5. Existence of real and quaternionic structures

Let $\lambda \in \Lambda_+$. A real or quaternionic structure exists on X_λ if and only if $\lambda \in \mathrm{Fix}(\Gamma, \Lambda_+)$. There is a map

$$ \beta : \mathrm{Fix}(\Gamma, \Lambda_+) \longrightarrow \mathrm{Br}(\mathbb{R}) $$

from the Γ-fixed dominant weights into the Brauer group $\mathrm{Br}(\mathbb{R}) = \{\mathbb{R}, \mathbb{H}\} \cong \mathbb{Z}/2$ such that X_λ admits a real (resp. a quaternionic) structure if and only if $\beta(\lambda) = \mathbb{R}$ (resp. $\beta(\lambda) = \mathbb{H}$). This map is additive in the sense that $\beta(\lambda + \mu) = \beta(\lambda)\beta(\mu)$.

For an arbitrary Γ-invariant dominant weight $\lambda = \bar{\lambda}$, the value $\beta(\lambda)$ can be determined as follows. Let $\lambda = \sum n_i \lambda_i \in \mathrm{Fix}(\Gamma, \Lambda_+)$ and put $Q(\lambda) = \{i \mid \lambda_i = \bar{\lambda}_i,\ \beta(\lambda_i) = \mathbb{H}\}$. Then $\beta(\lambda) = \mathbb{H}$ if and only if $\sum_{i \in Q(\lambda)} n_i \equiv 1 \pmod 2$. Thus, it suffices to know $\beta(\lambda_i)$ for the Γ-invariant fundamental weights.

4.B. Subgroups of classical groups

We are not really interested in homomorphism of G into the linear group $\mathrm{GL}_D(X)$, but rather into certain compact subgroups of $\mathrm{GL}_D(X)$. The Cartan decomposition of the groups $\mathrm{GL}_n\mathbb{R}$, $\mathrm{GL}_n\mathbb{C}$ and $\mathrm{GL}_n\mathbb{H}$ leads to the following result.

LEMMA 4.6. *Let (G,K) denote one of the pairs $(\mathrm{GL}_n\mathbb{R},\mathrm{O}(n))$, $(\mathrm{GL}_n\mathbb{C},\mathrm{U}(n))$, or $(\mathrm{GL}_n\mathbb{H},\mathrm{Sp}(n))$. Let*

$$\rho,\sigma : H \longrightarrow K \subseteq G$$

be homomorphisms which are conjugate by an element $g\in G$,

$$g\rho(h)=\sigma(h)g$$

for all $h\in H$. Then there exists an element $k\in K$ such that

$$k\rho(h)=\sigma(h)k.$$

PROOF. We use the Cartan decomposition $G=KP$. Note that P is a K-invariant subset of G. We decompose $g=kp$ with $k\in K$, $p\in P$. Then

$$\underbrace{\sigma(h)k}_{\in K}p = kp\rho(h) = \underbrace{k\rho(h)}_{\in K}\underbrace{\rho(h)^{-1}p\rho(h)}_{\in P}.$$

It follows from the uniqueness of the Cartan decomposition that

$$\sigma(h)k=k\rho(h) \text{ and } p=\rho(h)^{-1}p\rho(h).$$

□

This can be improved.

LEMMA 4.7. *Let $\rho,\sigma : H\longrightarrow \mathrm{SU}(n)$ be homomorphisms. If ρ and σ are conjugate by an element $u\in\mathrm{U}(n)$, then they are conjugate by an element in $\mathrm{SU}(n)$.*

Similarly, suppose that $\rho,\sigma : H\longrightarrow\mathrm{SO}(n)$ are homomorphisms which are conjugate by an element $u\in\mathrm{O}(n)$. If n is odd, then they are conjugate by an element of $\mathrm{SO}(n)$.

PROOF. We may multiply u by a number z such that $z^n=\det(u)^{-1}$. Then $\det(zu)=1$. □

This has the following consequence. In order to determine all subgroups of a given type of a compact classical group G, *up to automorphisms of G*, it suffices to classify all representations of groups of this given type on the natural G-module. If $G=\mathrm{SO}(n)$ is an orthogonal group, then the same method gives all subgroups of a given type in the universal covering $\mathrm{Spin}(n)$ (by considering the connected component of the lifted subgroup). The only thing that one has to check in this case is whether two different subgroups of $\mathrm{SO}(n)$ become equivalent in $\mathrm{Spin}(n)$. This happens indeed: for example, all subgroups of type $\mathfrak{b}_3$ in $\mathrm{Spin}(8)$ are equivalent under the automorphism group of $\mathrm{Spin}(8)$, although there are different (not quasi-isomorphic) representations of $\mathrm{Spin}(7)$ on $\mathbb{R}^8$.

4.C. Useful formulas

The results in this chapter follow from the tables in Tits [**104**], cp. also Onishchik-Vinberg [**81**] in particular p. 299–305, Salzmann *et al.* [**85**] p. 616–630. Bödi-Joswig [**6**] contains an algorithm which determines all irreducible representations of a given simple group (up to a certain dimension). I frequently used the computer implementation of this algorithm by the authors.

We use the following fact. Let $\lambda,\mu\in\Lambda_+$ be dominant weights of an almost simple compact Lie group. Suppose that

$$\mu=\lambda+n_1\lambda_1+\ldots+n_k\lambda_k,$$

with $n_1, \dots, n_k \geq 0$, and that not all n_i are zero. Then

$$\dim(\rho_\lambda) < \dim(\rho_\mu).$$

Note also that

$$\dim_{\mathbb{R}}({}^{\mathbb{R}}\rho) \in \{\dim_{\mathbb{C}}(\rho), 2\dim_{\mathbb{C}}(\rho))\}$$

and that

$$\dim_{\mathbb{H}}({}^{\mathbb{H}}\rho) \in \left\{\frac{1}{2}\dim_{\mathbb{C}}(\rho), \dim_{\mathbb{C}}(\rho)\right\}.$$

Suppose that G is simply connected. There exist homomorphisms $e_i : \mathrm{Cen}(G) \longrightarrow \mathbb{C}^*$ such that

$$\rho_{m_1\lambda_1+\cdots m_k\lambda_k}(z) = e_1(z)^{m_1} \cdots e_k(z)^{m_k}$$

for all $z \in \mathrm{Cen}(G)$. Thus, it suffices to know the e_i for $i = 1, \dots, k$ to determine the kernel of a representation. Note also that $\ker(\rho) = \ker({}^{\mathbb{R}}\rho) = \ker({}^{\mathbb{H}}\rho)$. We denote the kth exterior power of a vector space X by

$$\bigwedge^k X,$$

and the kth symmetric power by

$$\mathrm{S}^k X.$$

4.8. Fundamental weights for SU(2)

The weight lattice has one generator, λ_1, and $\bar{\lambda}_1 = \lambda_1$. The center is

$$\mathrm{Cen}(\mathrm{SU}(2)) \cong \langle z |\ z^2 = 1 \rangle \cong \mathbb{Z}/2$$

and $e_1(z) = -1$. Moreover, $\beta(\lambda_1) = \mathbb{H}$, hence

$$\beta(k\lambda_1) = \begin{cases} \mathbb{R} & \text{if } k \equiv 0 \pmod 2 \\ \mathbb{H} & \text{if } k \equiv 1 \pmod 2. \end{cases}$$

The dimension of $\rho_{k\lambda_1}$ is

$$\dim(\rho_{k\lambda_1}) = k + 1$$

and thus

$$\dim({}^{\mathbb{R}}\rho_{k\lambda_1}) = \begin{cases} k+1 & \text{if } k \equiv 0 \pmod 2 \\ 2(k+1) & \text{if } k \equiv 1 \pmod 2. \end{cases}$$

In fact

$$X_{k\lambda_1} = \mathrm{S}^k\mathbb{C}^2.$$

4.9. Fundamental weights for $\mathrm{SU}(n+1)$

We label the Dynkin diagram as follows, for $n \geq 2$.

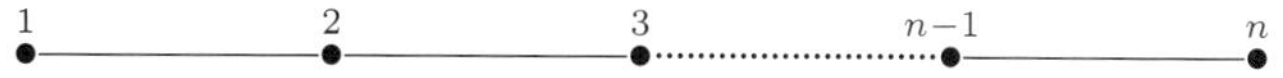

The center of the universal covering is

$$\mathrm{Cen}(\mathrm{SU}(n+1)) \cong \langle z |\ z^{n+1} = 1\rangle \cong \mathbb{Z}/(n+1),$$

and $e_i(z) = z^i$. The dimensions of the fundamental representations are

$$\dim(\rho_{\lambda_i}) = \binom{n+1}{i}$$

for $i = 1, \ldots, n$. The natural module is $R = X_{\lambda_1} = \mathbb{C}^{n+1}$, and

$$X_{\lambda_i} = \bigwedge^i R \qquad X_{k\lambda_1} = \mathrm{S}^k R \qquad X_{\lambda_1+\lambda_n} = \mathrm{Ad}.$$

The Galois group Γ acts as $\bar{\lambda}_i = \lambda_{n+1-i}$ for $i = 1, \ldots, n$. Moreover

$$\beta(\lambda_{\frac{n+1}{2}}) = \mathbb{R} \text{ for } n+1 \equiv 0 \pmod 2.$$

4.10. Low-dimensional simple $\mathrm{SU}(n+1)$-modules

Let X be a simple $\mathrm{SU}(n+1)$-$\mathbb{C}$-module, with $\dim(X) \leq 4(n+1)$, and let $G \subseteq \mathrm{GL}(X)$ denote the represented group. If $n \geq 9$, then (G, X) is quasi-isomorphic to $(\mathrm{SU}(n+1), \mathbb{C}^{n+1})$. For $2 \leq n \leq 8$, there are other low-dimensional simple modules.

G	weight	X	$\dim(X)$	${}^{\mathbb{R}}X$	$\dim({}^{\mathbb{R}}X)$
$\mathrm{SU}(n+1)$ $n \geq 2$	λ_1 or λ_n	$\mathbb{C}^{n+1}$	$n+1$	$\mathbb{C}^{n+1}$	$2(n+1)$
$\mathrm{SU}(9)$	λ_2 or λ_7	$\bigwedge^2 \mathbb{C}^9$	36	$\bigwedge^2 \mathbb{C}^9$	72
$\mathrm{SU}(8)/\langle z^4\rangle$	λ_2 or λ_6	$\bigwedge^2 \mathbb{C}^8$	28	$\bigwedge^2 \mathbb{C}^8$	56
$\mathrm{SU}(7)$	λ_2 or λ_5	$\bigwedge^2 \mathbb{C}^7$	21	$\bigwedge^2 \mathbb{C}^7$	42
$\mathrm{PSU}(7)$	$\lambda_1 + \lambda_6$		28		56
$\mathrm{SU}(6)/\langle z^3\rangle$	λ_2 or λ_4	$\bigwedge^2 \mathbb{C}^6$	15	$\bigwedge^2 \mathbb{C}^6$	30
$\mathrm{SU}(6)/\langle z^2\rangle$	λ_3	$\bigwedge^3 \mathbb{C}^6$	20	$\bigwedge^3 \mathbb{C}^6$	40
$\mathrm{SU}(6)/\langle z^3\rangle$	$2\lambda_1$ or $2\lambda_5$	$\mathrm{S}^2\mathbb{C}^6$	21	$\mathrm{S}^2\mathbb{C}^6$	42
$\mathrm{SU}(5)$	λ_2 or λ_3	$\bigwedge^2 \mathbb{C}^5$	10	$\bigwedge^2 \mathbb{C}^5$	20
$\mathrm{SU}(5)$	$2\lambda_1$ or $2\lambda_3$		15		30
$\mathrm{SO}(6)$	λ_2	$\mathbb{C}^6$	6	$\mathbb{R}^6$	6
$\mathrm{PSU}(4)$	$\lambda_1 + \lambda_3$	$\mathfrak{sl}_4\mathbb{C}$	15	$\mathfrak{su}_4\mathbb{C}$	15
$\mathrm{SO}(6)$	$2\lambda_1$ or $2\lambda_3$	$\mathrm{S}^2\mathbb{C}^4$	10	$\mathrm{S}^2\mathbb{C}^4$	20
$\mathrm{SU}(3)$	$2\lambda_1$ or $2\lambda_2$	$\mathrm{S}^2\mathbb{C}^3$	6	$\mathrm{S}^2\mathbb{C}^3$	12
$\mathrm{PSU}(3)$	$\lambda_1 + \lambda_2$	$\mathfrak{sl}_3\mathbb{C}$	8	$\mathfrak{su}_3\mathbb{C}$	8
$\mathrm{SU}(3)$	$3\lambda_1$ or $3\lambda_2$	$\mathrm{S}^3\mathbb{C}^3$	10	$\mathrm{S}^3\mathbb{C}^3$	20

Proof. We have $\dim(\rho_{\lambda_2}) = n(n+1)/2 > 4n+4$ for $n \geq 9$. Now $\dim(\rho_{\lambda_i}) \geq \dim(\rho_{\lambda_2})$ for $i = 2, \ldots, n-2$, and $\dim(\rho_{2\lambda_1}) = \binom{n+2}{2} > 4(n+1)$ for $n \geq 7$. Therefore $\dim(\rho_\lambda) > \dim(\rho_{\lambda_1})$ for $n \geq 9$ and $\lambda \neq \lambda_1, \lambda_n$. □

4.11. FUNDAMENTAL WEIGHTS FOR $\mathrm{Spin}(2n+1)$

We label the Dynkin diagram as follows, for $n \geq 3$.

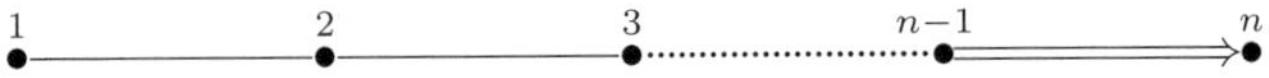

The center of the universal covering is

$$\mathrm{Cen}(\mathrm{Spin}(2n+1)) = \langle z |\ z^2 = 1 \rangle \cong \mathbb{Z}/2,$$

$e_i(z) = 1$ for $i = 1, \ldots, n-1$, and $e_n(z) = -1$. The dimensions of the fundamental representations are

$$\dim(\rho_{\lambda_i}) = \binom{2n+1}{i}$$

for $i = 1, \ldots, n-1$, and

$$\dim(\rho_{\lambda_n}) = 2^n.$$

The natural module is $R = X_{\lambda_1} = \mathbb{C}^{2n+1}$, and

$$X_{\lambda_i} = \bigwedge^i R \quad (1 \leq i \leq n-1) \qquad X_{2\lambda_n} = \bigwedge^n R \qquad X_{\lambda_2} = \bigwedge^2 R = \mathrm{Ad} \text{ if } n \geq 3,$$

$\mathrm{Ad} = X_{2\lambda_2}$ for $n = 2$. The Galois group Γ acts as $\lambda_i = \bar{\lambda}_i$ for $i = 1, \ldots, n$. Moreover

$$\beta(\lambda_i) = \mathbb{R} \text{ for } i = 1, \ldots, n-1$$

and

$$\beta(\lambda_n) = \begin{cases} \mathbb{R} & \text{for } n \equiv 0, 3 \pmod 4 \\ \mathbb{H} & \text{for } n \equiv 1, 2 \pmod 4. \end{cases}$$

4.12. LOW-DIMENSIONAL SIMPLE $\mathrm{Spin}(2n+1)$-MODULES

Let X be a simple $\mathrm{Spin}(2n+1)$-$\mathbb{C}$-module, with $\dim(X) \leq 4(2n+1)$, and let $G \subseteq \mathrm{GL}(X)$ denote the represented group. If $n \geq 6$, then (G, X) is quasi-isomorphic to $(\mathrm{SO}(2n+1), \mathbb{C}^{2n+1})$. For $2 \leq n \leq 5$, we have additional simple modules.

G	weight	X	$\dim(X)$	${}^{\mathbb{R}}X$	$\dim({}^{\mathbb{R}}X)$
$\mathrm{SO}(2n+1)$ $n \geq 2$	λ_1	$\mathbb{C}^{2n+1}$	$2n+1$	$\mathbb{R}^{2n+1}$	$2n+1$
$\mathrm{Spin}(11)$	λ_5		32		64
$\mathrm{Spin}(9)$	λ_4	$\mathbb{O}^2 \otimes_{\mathbb{R}} \mathbb{C}$	16	$\mathbb{O}^2$	16
$\mathrm{SO}(9)$	λ_2	$\mathfrak{so}_9\mathbb{C}$	36	$\mathfrak{so}_9$	36
$\mathrm{Spin}(7)$	λ_3	$\mathbb{C}^8$	8	$\mathbb{R}^8$	8
$\mathrm{SO}(7)$	λ_2	$\mathfrak{so}_7\mathbb{C}$	21	$\mathfrak{so}_7$	21
$\mathrm{SO}(7)$	$2\lambda_1$		27		27
$\mathrm{Sp}(2)$	λ_2	$\mathbb{H}^2$	4	$\mathbb{H}^2$	8
$\mathrm{SO}(5)$	$2\lambda_2$	$\mathfrak{so}_5\mathbb{C}$	10	$\mathfrak{so}_5$	10
$\mathrm{SO}(5)$	$2\lambda_1$		14		14
$\mathrm{SO}(5)$	$\lambda_1 + \lambda_2$		16		32
$\mathrm{SO}(5)$	$3\lambda_1$		20		40

PROOF. We have $\dim(\rho_{\lambda_2}) = n(2n+1) > 4(2n+1)$ for $n \geq 5$, and $\dim(\rho_{\lambda_i}) > \dim(\rho_{\lambda_2})$ for $i = 1, \ldots, n-1$. Moreover, $2^n > 4(2n+1)$ for $n \geq 6$. □

4.13. FUNDAMENTAL WEIGHTS FOR $\mathrm{Sp}(n)$

We label the Dynkin diagram as follows, for $n \geq 3$.

The center of the universal covering is

$$\mathrm{Cen}(\mathrm{Sp}(n)) = \langle z | \ z^2 = 1 \rangle \cong \mathbb{Z}/2,$$

and $e_i(z) = (-1)^i$. The dimensions of the fundamental representations are

$$\dim(\rho_1) = 2n$$

and

$$\dim(\rho_{\lambda_i}) = \binom{2n}{i} - \binom{2n}{i-2} = \binom{2n}{i-2} \frac{(2n+1)(2n-2i+2)}{(2n-i+1)(2n-i+2)}$$

for $i = 2, \ldots, n$. The natural module is $R = X_{\lambda_1} = \mathbb{C}^{2n} = \mathbb{H}^n$, and

$$X_{k\lambda_1} = \mathrm{S}^k R \qquad X_{2\lambda_1} = \mathrm{Ad}.$$

The Galois group Γ acts as $\lambda_i = \bar{\lambda}_i$ for $i = 1, \ldots, n$. Moreover

$$\beta(\lambda_i) = \begin{cases} \mathbb{R} & \text{for } i \equiv 0 \pmod 2 \\ \mathbb{H} & \text{for } i \equiv 1 \pmod 2 \end{cases}$$

4.14. LOW-DIMENSIONAL SIMPLE $\mathrm{Sp}(n)$-MODULES

Let X be a simple $\mathrm{Sp}(n)$-$\mathbb{C}$-module, with $\dim(X) \leq 4 \cdot 2n$, and let $G \subseteq \mathrm{GL}(X)$ denote the represented group. If $n \geq 5$, then (G, X) is quasi-isomorphic to $(\mathrm{Sp}(n), \mathbb{H}^n)$. For $2 \leq n \leq 4$, we have additional simple modules.

G	weight	X	$\dim(X)$	${}^{\mathbb{R}}X$	$\dim({}^{\mathbb{R}}X)$	${}^{\mathbb{H}}X$	$\dim({}^{\mathbb{H}}X)$
$\mathrm{Sp}(n)$ $n \geq 3$	λ_1	$\mathbb{H}^n$	$2n$	$\mathbb{H}^n$	$4n$	$\mathbb{H}^n$	n
$\mathrm{PSp}(4)$	λ_2		27		27		27
$\mathrm{PSp}(3)$	λ_2		14		14		14
$\mathrm{Sp}(3)$	λ_3		14		28		7
$\mathrm{PSp}(3)$	$2\lambda_1$	$\mathfrak{sp}_6\mathbb{C}$	21	$\mathfrak{sp}_3$	21		21

PROOF. We have $\dim(\rho_{\lambda_2}) = \binom{2n}{2} - 1 > 8n$ for $n \geq 5$. An unpleasant calculation shows that $\dim(\rho_{\lambda_k}) \geq \dim(\rho_{\lambda_2})$ for $k = 3, \ldots, n$. $\square$

4.15. Fundamental weights for Spin($2n$)

We label the Dynkin diagram as follows, for $n \geq 4$.

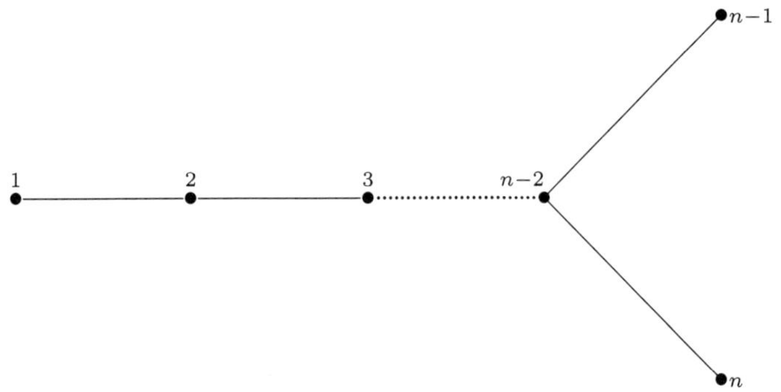

The center of the universal covering is as follows.

If n is *odd*, then

$$\mathrm{Cen}(\mathrm{Spin}(2n)) = \langle z \mid z^4 = 1 \rangle \cong \mathbb{Z}/4,$$

$e_i(z) = (-1)^i$ for $i = 1, \dots, n-2$, and $e_{n-1}(z) = e_n(z) = \sqrt{-1}$.

If n is *even*, then

$$\mathrm{Cen}(\mathrm{Spin}(2n)) = \langle z, z' \mid z^2 = z'^2 = 1,\ zz' = z'z \rangle \cong \mathbb{Z}/2 \oplus \mathbb{Z}/2,$$

$e_i(z) = e_i(z') = (-1)^i$ for $i = 1, \dots, n-2$, and $e_{n-1}(z) = e_n(z') = 1$, $e_{n-1}(z') = e_n(z) = -1$.

The dimensions of the fundamental representations are

$$\dim(\rho_{\lambda_i}) = \binom{2n}{i}$$

for $i = 1, \dots, n-2$, and

$$\dim(\rho_{\lambda_{n-1}}) = \dim(\rho_{\lambda_n}) = 2^{n-1}.$$

The natural module is $R = X_{\lambda_1} = \mathbb{C}^{2n}$, and

$$X_{\lambda_i} = \bigwedge^i R \quad (1 \leq i \leq n-2) \qquad X_{\lambda_2} = \bigwedge^2 R = \mathrm{Ad},$$

$X_{\lambda_{n-1}+\lambda_n} = \bigwedge^{n-1} R$. The Galois group Γ acts as $\lambda_i = \bar{\lambda}_i$ for $i = 1, \dots, n-2$, and

$$\bar{\lambda}_{n-1} = \begin{cases} \lambda_n & \text{for } n \equiv 1 \pmod 2 \\ \lambda_{n-1} & \text{for } n \equiv 0 \pmod 2. \end{cases}$$

Moreover

$$\beta(\lambda_i) = \mathbb{R}$$

for $i - 1, \dots, n-2$, and

$$\beta(\lambda_{n-1}) = \beta(\lambda_n) = \begin{cases} \mathbb{R} & \text{for } n \equiv 0 \pmod 4 \\ \mathbb{H} & \text{for } n \equiv 2 \pmod 4. \end{cases}$$

4.16. Low-dimensional simple Spin($2n$)-modules

Let X be a simple Spin($2n$)-$\mathbb{C}$-module, with $\dim(X) \leq 4 \cdot 2n$, and let $G \subseteq \mathrm{GL}(X)$ denote the represented group. If $n \geq 7$, then (G, X) is quasi-isomorphic to $(\mathrm{SO}(2n), \mathbb{C}^n)$. For $n = 4, 5, 6$ there are additional simple modules.

G	weight	X	$\dim(X)$	${}^{\mathbb{R}}X$	$\dim({}^{\mathbb{R}}X)$
$\mathrm{SO}(2n)$ $n \geq 5$	λ_1	$\mathbb{C}^{2n}$	$2n$	$\mathbb{R}^{2n}$	$2n$
$\mathrm{Spin}(12)$	λ_5 or λ_6	$\mathbb{H}^{16}$	16	$\mathbb{H}^{16}$	64
$\mathrm{Spin}(10)$	λ_4 or λ_5	$\mathbb{C}^{16}$	16	$\mathbb{C}^{16}$	32
$\mathrm{SO}(8)$	λ_1, λ_3 or λ_4	$\mathbb{C}^8$	8	$\mathbb{R}^8$	8
$\mathrm{PSO}(8)$	λ_2	$\mathfrak{so}_8\mathbb{C}$	28	$\mathfrak{so}_8$	28

PROOF. We have $\dim(\rho_{\lambda_2}) = \binom{2n}{2} > 8n$ for $n \geq 5$, and $\dim(\rho_{\lambda_i}) > \dim(\rho_{\lambda_2})$ for $i = 3, \ldots, n-2$. Moreover, $2^{n-1} > 8n$ for $n \geq 7$. $\square$

4.17. Fundamental weights for E_6

We label the Dynkin diagram as follows.

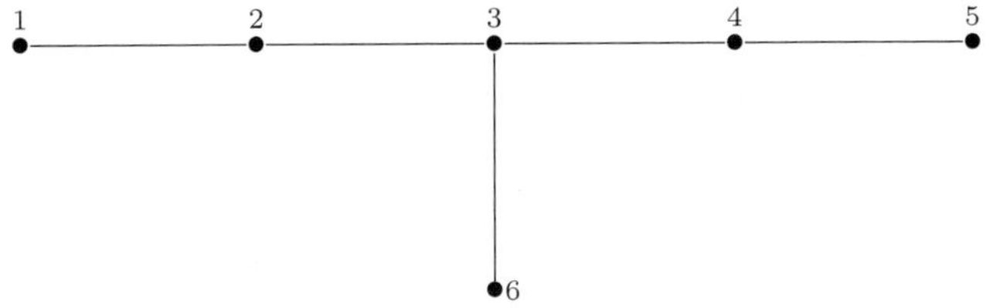

The center of the universal covering is

$$\mathrm{Cen}(E_6) = \langle z | \; z^3 = 1 \rangle \cong \mathbb{Z}/3,$$

and $e_i(z) = \exp(2\pi\sqrt{-1}/3)$. The dimensions of the fundamental representations are

$$\dim(\rho_{\lambda_i}) = \begin{cases} 27 & \text{for } i = 1, 5 \\ 351 & \text{for } i = 2, 4 \\ 2\,925 & \text{for } i = 3 \\ 78 & \text{for } i = 6. \end{cases}$$

The natural module is X_{λ_1}, and $X_{\lambda_6} = \mathrm{Ad}$.

The Galois group Γ acts as $\lambda_i = \bar{\lambda}_{6-i}$ for $i = 1, 2$, $\bar{\lambda}_3 = \lambda_3$, and $\bar{\lambda}_6 = \lambda_6$. Moreover, $\beta(\lambda_3) = \beta(\lambda_6) = \mathbb{R}$.

4.18. Low-dimensional simple E_6-modules

The E_6-modules of dimension at most 108 are the following.

G	weight	X	$\dim(X)$	${}^{\mathbb{R}}X$	$\dim({}^{\mathbb{R}}X)$
E_6	λ_1 or λ_5		27		54
PE_6	λ_6	$\mathfrak{e}_6\mathbb{C}$	78	$\mathfrak{e}_6$	78

□

4.19. FUNDAMENTAL WEIGHTS FOR E_7

We label the Dynkin diagram as follows.

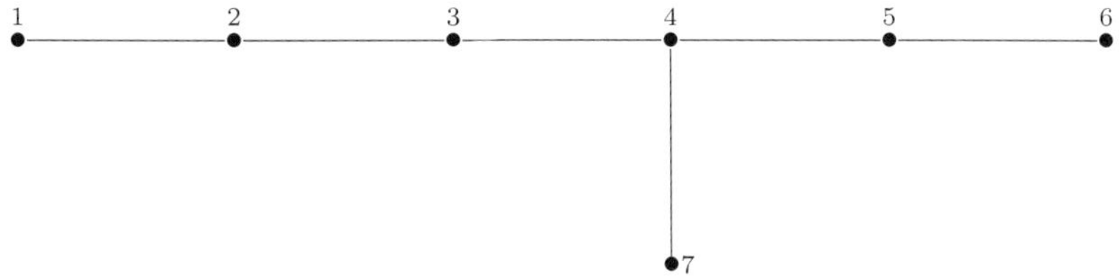

The center of the universal covering is

$$\mathrm{Cen}(\mathrm{E}_7) = \langle z \mid z^2 = 1 \rangle \cong \mathbb{Z}/2,$$

and $e_1(z) = e_3(z) = e_7(z) = -1$, $e_2(z) = e_4(z) = e_5(z) = e_6(z) = 1$. The dimensions of the fundamental representations are

$$\dim(\rho_{\lambda_i}) = \begin{cases} 56 & \text{for } i = 1 \\ 1\,539 & \text{for } i = 2 \\ 27\,664 & \text{for } i = 3 \\ 365\,750 & \text{for } i = 4 \\ 8\,645 & \text{for } i = 5 \\ 133 & \text{for } i = 6 \\ 912 & \text{for } i = 7. \end{cases}$$

The natural module is X_{λ_1}, and $X_{\lambda_6} = \mathrm{Ad}$.

The Galois group Γ acts as $\lambda_i = \bar{\lambda}_i$ for $i = 1, \dots, 7$, and

$$\beta(\lambda_i) = \begin{cases} \mathbb{R} & \text{for } i = 2, 4, 5, 6 \\ \mathbb{H} & \text{for } i = 1, 3, 7. \end{cases}$$

4.20. LOW-DIMENSIONAL SIMPLE E_7-MODULES

The E_6-modules of dimension at most 224 are the following.

G	weight	X	$\dim(x)$	${}^{\mathbb{R}}X$	$\dim({}^{\mathbb{R}}X)$
E_7	λ_1		56		112
PE_7	λ_6	$\mathfrak{e}_7\mathbb{C}$	133	$\mathfrak{e}_7$	133

□

4.21. Fundamental weights for E_8

We label the Dynkin diagram as follows.

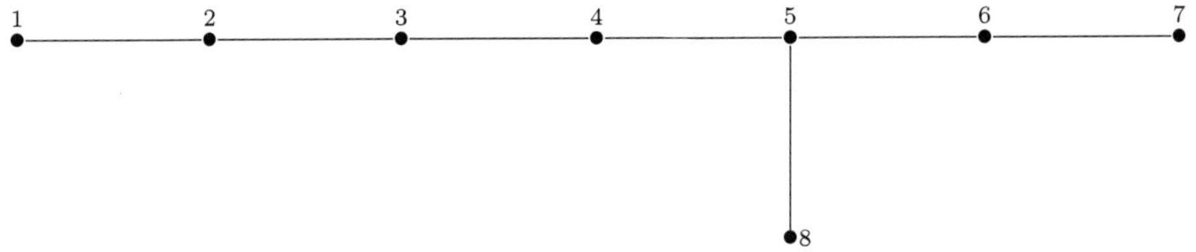

The group is simply connected and simple. The dimensions of the fundamental representations are

$$\dim(\rho_{\lambda_i}) = \begin{cases} 248 & \text{for } i = 1 \\ 30\,380 & \text{for } i = 2 \\ 2\,450\,240 & \text{for } i = 3 \\ 146\,325\,270 & \text{for } i = 4 \\ 6\,899\,079\,264 & \text{for } i = 5 \\ 6\,696\,000 & \text{for } i = 6 \\ 3\,875 & \text{for } i = 7 \\ 147\,250 & \text{for } i = 8. \end{cases}$$

The natural module is $X_{\lambda_1} = \mathrm{Ad}$.

The Galois group acts as $\lambda_i = \bar{\lambda}_i$ for $i = 1, \ldots, 8$, and $\beta(\lambda_i) = \mathbb{R}$ for $i = 1, \ldots, 8$.

4.22. Low-dimensional simple E_8-modules

The E_6-modules of dimension at most 992 are the following.

G	weight	X	$\dim(X)$	${}^{\mathbb{R}}X$	$\dim({}^{\mathbb{R}}X)$
E_8	λ_1	$\mathfrak{e}_8\mathbb{C}$	248	$\mathfrak{e}_8$	248

□

4.23. FUNDAMENTAL WEIGHTS FOR F_4

We label the Dynkin diagram as follows.

The group is simple and simply connected. The dimensions of the fundamental representations are

$$\dim(\rho_{\lambda_i}) = \begin{cases} 26 & \text{for } i = 1 \\ 273 & \text{for } i = 2 \\ 1\,274 & \text{for } i = 3 \\ 52 & \text{for } i = 4. \end{cases}$$

The natural module is X_{λ_1}, and $X_{\lambda_4} = \mathrm{Ad}$. The Galois group Γ acts as $\lambda_i = \bar{\lambda}_i$, and $\beta(\lambda_i) = \mathbb{R}$ for $i = 1, \dots, 4$.

4.24. LOW-DIMENSIONAL SIMPLE F_4-MODULES

The F_4-modules of dimension at most 104 are the following.

G	weight	X	$\dim(X)$	${}^{\mathbb{R}}X$	$\dim({}^{\mathbb{R}}X)$
F_4	λ_1		26		26
F_4	λ_4	$\mathfrak{f}_4\mathbb{C}$	52	$\mathfrak{f}_4$	52

□

4.25. Fundamental weights for G_2

We label the Dynkin diagram as follows.

The group is simple and simply connected. The dimensions of the fundamental representations are

$$\dim(\rho_{\lambda_1}) = 7$$

and

$$\dim(\rho_{\lambda_2}) = 14.$$

The natural module is X_{λ_1}, and $X_{\lambda_2} = \mathrm{Ad}$. The Galois group acts as $\lambda_i = \bar{\lambda}_i$ for $i = 1, 2$, and

$$\beta(\lambda_i) = \mathbb{R}$$

for $i = 1, 2$.

4.26. Low-dimensional simple G_2-modules

The G_2-modules of dimension at most 28 are the following.

G	weight	X	$\dim(X)$	${}^{\mathbb{R}}X$	$\dim({}^{\mathbb{R}}X)$
G_2	λ_1	$\mathbb{C}^7$	7	$\mathbb{R}^7$	7
G_2	λ_2	$\mathfrak{g}_2\mathbb{C}$	14	$\mathfrak{g}_2$	14
G_2	$2\lambda_1$		27		27

□

4.27. Certain subgroups of exceptional Lie groups

Borel-De Siebenthal [**13**] determined the maximal connected subgroups of maximal rank for all compact almost simple Lie groups. Dynkin [**29**] and Seitz [**89**] determined the maximal connected subgroups of the exceptional complex almost simple Lie groups which have strictly smaller rank. The complexification is a functorial equivalence between compact connected Lie groups and reductive algebraic $\mathbb{C}$-groups. Therefore, the results of Dynkin and Seitz apply to compact almost simple Lie groups. We obtain in particular the following result: The two standard inclusions

$$\mathrm{G}_2 \subseteq \mathrm{F}_4 \text{ and } \mathrm{F}_4 \subseteq \mathrm{E}_6$$

are unique up to conjugation, cp. Seitz [**89**] Table 1, p. 193.

The reasoning here is as follows. Let $\mathrm{G}_2^{\mathbb{C}}$ and $\mathrm{F}_4^{\mathbb{C}}$ denote the corresponding complex Lie groups. An inclusion $\mathrm{G}_2 \hookrightarrow \mathrm{F}_4$ yields (via complexification) an inclusion $\mathrm{G}_2^{\mathbb{C}} \xhookrightarrow{\phi} \mathrm{F}_4^{\mathbb{C}}$. By the result of Dynkin and Seitz, this inclusion is conjugate to the standard inclusion $\mathrm{G}_2^{\mathbb{C}} \xhookrightarrow{s} \mathrm{F}_4^{\mathbb{C}}$ by an element $g \in \mathrm{F}_4^{\mathbb{C}}$. Since $\mathrm{G}_2 \subseteq \mathrm{G}_2^{\mathbb{C}}$ is a maximal compact subgroup, we can assume that g transforms $\phi(\mathrm{G}_2)$ into $s(\mathrm{G}_2)$ (all maximal compact subgroups in $\mathrm{G}_2^{\mathbb{C}}$ are conjugate). Now write $g = kp$ as in 4.6, with $k \in \mathrm{F}_4$. It follows by a similar argument as in 4.6 that $\phi(\mathrm{G}_2)$ and $s(\mathrm{G}_2)$ are conjugate under $k \in \mathrm{F}_4$.

The maximal connected subgroups of rank 2 in G_2 are according to Borel-De Siebenthal [**13**] the groups SO(4) (the stabilizer of the quaternion algebra $\mathbb{H} \subseteq \mathbb{O}$), and SU(3) (the elementwise stabilizer of $\mathbb{C} \subseteq \mathbb{O}$). According to Dynkin [**29**] and Seitz [**89**] the representation ${}^{\mathbb{R}}\rho_{6\lambda_1}$ lifts to a maximal connected subgroup of G_2. The following table gives all subgroups H of type $\mathfrak{a}_1$ in G_2.

3-dimensional connected subgroups of G_2

H	ρ	maximal subgroup containing H
SO(3)	$2 \cdot {}^{\mathbb{R}}\rho_{2\lambda_1}$	SU(3)
SU(2)	${}^{\mathbb{R}}\rho_{\lambda_1}$	SU(3)
SU(2)	${}^{\mathbb{R}}\rho_{\lambda_1} + {}^{\mathbb{R}}\rho_{2\lambda_1}$	SO(4)
SO(3)	${}^{\mathbb{R}}\rho_{6\lambda_1}$	

4.D. The Dynkin index

Let $\phi : H \longrightarrow G$ be a homomorphism between compact almost simple Lie groups. There is a corresponding homomorphism

$$\pi_3(\phi) : \pi_3(H) \longrightarrow \pi_3(G).$$

Both groups are infinite cyclic, hence there exists a number $j = j_\phi \in \mathbb{N}$ such that the cokernel of $\pi_3(\phi)$ is cyclic of order j. This number is called the *Dynkin index* of ϕ. It is clear that the Dynkin index is multiplicative,

$$j_{\phi \circ \psi} = j_\phi \cdot j_\psi.$$

Moreover, if ρ and σ are representations, then

$$j_{\rho+\sigma} = j_\rho + j_\sigma,$$

cp. Onishchik [**80**] Ch. 5, Thm. 2, §17.2, p. 257. The following diagrams show the Dynkin indices of various natural inclusions between compact almost simple Lie groups. The groups on the right are the stable limits of the classical groups.

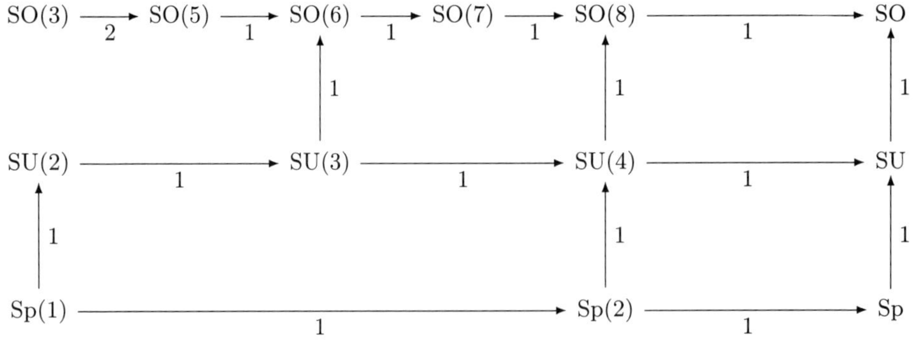

Here, one can fill in the following subdiagram.

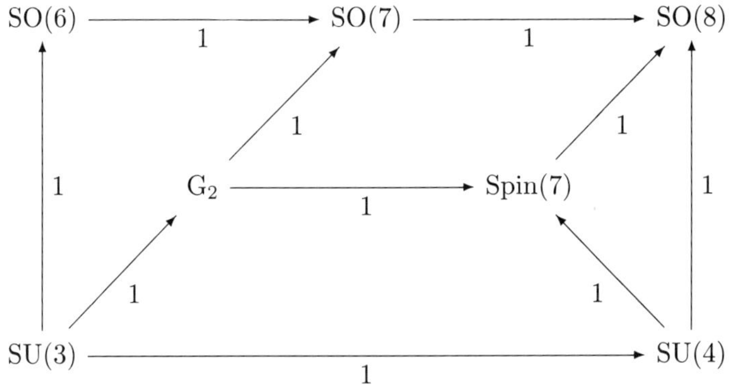

Finally, tensoring with $\mathbb{C}$ and $\mathbb{H}$ yields a diagram

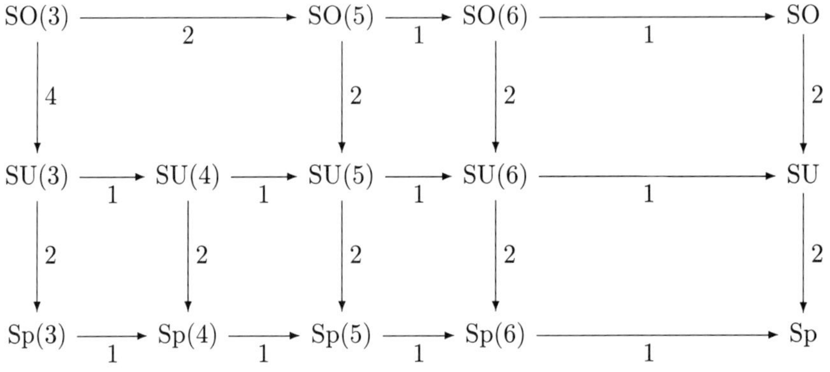

Besides this, we use the following result for representations of SU(2). The Dynkin index of $\rho_{k\lambda_1} : \mathrm{SU}(2) \longrightarrow \mathrm{SU}(k+1)$ is

$$j_{\rho_{k\lambda_1}} = \binom{k+2}{3},$$

cp. Onishchik [**80**] p. 61.

CHAPTER 5

The case when G is simple

In this chapter we classify all homogeneous spaces $X = G/H$ with the following properties.

(i) G/H is compact and 1-connected.
(ii) G is effective, compact and almost simple.
(iii) G/H has the same integral cohomology as a product $\mathbb{S}^{n_1} \times \mathbb{S}^{n_2}$, where $3 \leq n_1 \leq n_2$, and n_2 is odd.

There are two cases: either n_1 is odd, or n_1 is even.
Case (I): n_1 is odd, $\mathbf{H}^\bullet(G/H) \cong \bigwedge_{\mathbb{Z}}(u, v)$, with $\deg(u) = n_1$ and $\deg(v) = n_2$. In this situation Theorem 3.7 applies.
Case (II): n_1 is even, $\mathbf{H}^\bullet(G/H) \cong \mathbb{Z}[a]/(a^2) \otimes \bigwedge_{\mathbb{Z}}(u)$, with $\deg(a) = n_1$ and $\deg(u) = n_2$. In this situation Theorem 3.11 applies.

These two cases are considered in the next two sections. Each section ends with a complete list of all possible pairs (G, H), see 5.6, 5.14.

5.A. Case (I): $\mathbf{H}^\bullet(X) = \bigwedge_{\mathbb{Z}}(u, v)$.

If n_1 is odd, then $\mathbf{H}^\bullet(X; \mathbb{Q}) \cong \bigwedge_{\mathbb{Q}}(u, v)$. By Theorem 3.7 $\mathrm{rk}(G) - \mathrm{rk}(H) = 2$. Moreover, $P_H^1 \xleftarrow{\cong} \mathcal{P}_G^1 = 0$, and $P_H^3 \longleftarrow P_G^3$ is a surjection; thus, H is either trivial or almost simple. Let $(m_1^G, \ldots, m_n^G)$ denote the degrees of the homogeneous generators of G, and let $(m_1^H, \ldots, m_{n-2}^H)$ denote the corresponding numbers for H. After a suitable permutation of the indices, $m_i^G = m_i^H$ for $i = 1, \ldots, n-2$, cp. 3.8. We check the various possibilities.

5.1. G of type $\mathfrak{a}_n$

The homogeneous generators of the rational cohomology of $\mathrm{SU}(n+1)$ have degrees

$$(3, 5, \ldots, 2n+1).$$

If $n \geq 8$, then the only type of simple group which fits in is $\mathfrak{a}_{n-2}$; in low dimensions, we have also to consider the pairs $(\mathfrak{a}_7, \mathfrak{d}_5)$, $(\mathfrak{a}_5, \mathfrak{b}_3)$, $(\mathfrak{a}_5, \mathfrak{c}_3)$, and $(\mathfrak{a}_4, \mathfrak{c}_2)$.

$\boxed{(\mathfrak{a}_n, \mathfrak{a}_{n-2})}$ For $n \geq 4$, it follows from 4.10 that $(G, H) = (\mathrm{SU}(n+1), \mathrm{SU}(n-1))$.
Suppose that $n = 3$. Then H is of type $\mathfrak{a}_1$. By 4.8, the representations of $\mathrm{SU}(2)$ on $\mathbb{C}^4$ are $\mathrm{SU}(2)$ (i.e. ρ_{λ_1}), $\mathrm{SO}(3)$ (i.e. $\rho_{2\lambda_1}$), $2\rho_{\lambda_1}$, and $\rho_{3\lambda_1}$. Note that $\mathrm{SU}(4) \cong \mathrm{Spin}(6)$; the subgroup $2 \cdot \rho_{\lambda_1}(\mathrm{SU}(2)) \subseteq \mathrm{SU}(4)$ contains the center, and $\mathrm{SU}(4)/2 \cdot \rho_{\lambda_1}(\mathrm{SU}(2)) = \mathrm{SO}(6)/\mathrm{SO}(3) = V_3(\mathbb{R}^6)$.

$\boxed{(\mathfrak{a}_7, \mathfrak{d}_5) \text{ and } (\mathfrak{a}_5, \mathfrak{b}_3)}$ The group $\mathrm{Spin}(10)$ has no non-trivial representation on $\mathbb{C}^8$; similarly, there is no non-trivial representation of $\mathrm{Spin}(7)$ on $\mathbb{C}^6$, cp. 4.12. This shows that the pairs $(\mathfrak{a}_7, \mathfrak{d}_5)$ and $(\mathfrak{a}_5, \mathfrak{b}_3)$ do not exist.

$(\mathfrak{a}_5, \mathfrak{c}_3)$ and $(\mathfrak{a}_4, \mathfrak{c}_2)$ There is a (unique) representation of Sp(3) on $\mathbb{C}^6 = \mathbb{H}^3$, the natural one, and two representations of Sp(2) on $\mathbb{C}^5$, arising from the inclusions $\mathrm{SO}(5) \subseteq \mathrm{SU}(5)$ and $\mathrm{Sp}(2) \subseteq \mathrm{SU}(4) \subseteq \mathrm{SU}(5)$, cp. 4.12.

We obtain the following list of groups and homogeneous spaces. The structure of π_3 follows from the formula for the Dynkin index of the various representations of SU(2).

G of type $\mathrm{SU}(n+1)$

G	H	$\mathrm{Cen}_G(H)^\circ$	(n_1, n_2)	G/H	Remarks
$\mathrm{SU}(n+1)$ $n \geq 3$	$\mathrm{SU}(n-1)$	U(2)	$(2n-1, 2n+1)$	$V_2(\mathbb{C}^{n+1})$	
SU(6)	Sp(3)	1	(5, 9)		(see below)
SU(5)	Sp(2)	U(1)	(5, 9)		(see below)
SU(5)	SO(5)	1	(5, 9)		$\pi_3 = \mathbb{Z}/2$
SU(4)	SO(3)	U(1)	(5, 7)		$\pi_3 = \mathbb{Z}/4$
SU(4)	$\rho_{3\lambda_1}$	1	(5, 7)		$\pi_3 = \mathbb{Z}/10$
SO(6)	SO(3)	SO(3)	(5, 7)	$V_3(\mathbb{R}^6)$	$\pi_3 = \mathbb{Z}/2$
SU(3)	1	SU(3)	(3, 5)	$V_2(\mathbb{C}^3)$	

There is one coincidence in this table, arising from the inclusion $\mathrm{SU}(5) \subseteq \mathrm{SU}(6)$:

$$\mathrm{SU}(5)/\mathrm{Sp}(2) = \mathrm{SU}(6)/\mathrm{Sp}(3)$$

5.2. G OF TYPE $\mathfrak{b}_n$

The homogeneous generators of the rational cohomology of $\mathrm{Spin}(2n+1)$ have degrees

$$(3, 7, \ldots, 4n-1).$$

If $n \geq 10$, then the only types of simple groups which fit in are $\mathfrak{b}_{n-2}$ and $\mathfrak{c}_{n-2}$. However, $\mathbb{R}^{2n+1}$ is not a non-trivial $\mathrm{Sp}(n-2)$-module for $n \geq 5$, so $(\mathfrak{b}_n, \mathfrak{c}_{n-2})$ is excluded in this range, and $\mathfrak{b}_2 \cong \mathfrak{c}_2$ anyway. In low dimensions, we have also to consider the pairs $(\mathfrak{b}_9, \mathfrak{e}_7)$, $(\mathfrak{b}_6, \mathfrak{f}_4)$, and $(\mathfrak{b}_4, \mathfrak{g}_2)$.

$(\mathfrak{b}_n, \mathfrak{b}_{n-2})$ If $n \geq 6$, then $(G, H) = (\mathrm{SO}(2n+1), \mathrm{SO}(2n-3))$, cp. 4.12.

The group Spin(7) has two non-trivial representations on $\mathbb{R}^{11}$, corresponding to the standard inclusion $\mathrm{SO}(7) \subseteq \mathrm{SO}(9)$, and to the inclusion $\mathrm{Spin}(7) \subseteq \mathrm{SO}(8)$, cp. 4.12.

Similarly, Spin(5) has two non-trivial representations on $\mathbb{R}^9$, one arising from the inclusion $\mathrm{SO}(5) \subseteq \mathrm{SO}(9)$, and one arising from $\mathrm{Sp}(2) \subseteq \mathrm{SO}(8) \subseteq \mathrm{SO}(9)$.

The representations of SU(2) on $\mathbb{R}^7$ are SO(3) (i.e. ${}^{\mathbb{R}}\rho_{2\lambda_1}$), SU(2) (i.e. ${}^{\mathbb{R}}\rho_{\lambda_1}$), ${}^{\mathbb{R}}\rho_{\lambda_1} + {}^{\mathbb{R}}\rho_{2\lambda_1}$, ${}^{\mathbb{R}}\rho_{4\lambda_1}$, and ${}^{\mathbb{R}}\rho_{6\lambda_1}$. We denote the corresponding homomorphisms into Spin(7) by a tilde,

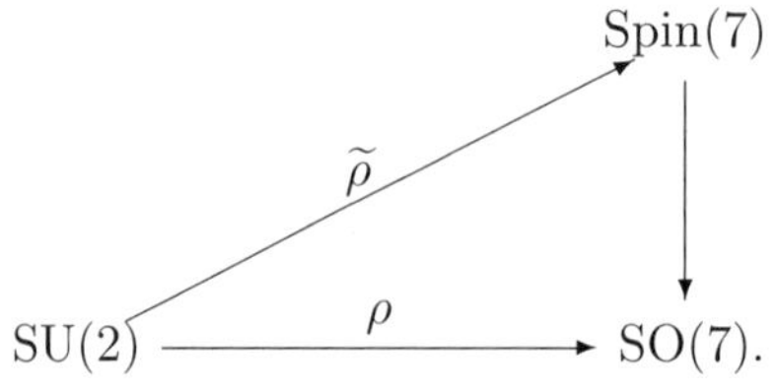

$(\mathfrak{b}_9, \mathfrak{e}_7)$ and $(\mathfrak{b}_6, \mathfrak{f}_4)$ The compact groups E_7 and F_4 do not have non-trivial 19- and 13-dimensional representations, respectively, cp. 4.24 and 4.20, thus the pairs $(\mathfrak{b}_9, \mathfrak{e}_7)$ and $(\mathfrak{b}_6, \mathfrak{f}_4)$ do not exist.

$(\mathfrak{b}_4, \mathfrak{g}_2)$ There is only one 8-dimensional representation of G_2 on $\mathbb{R}^8$, given by the action of G_2 on the Cayley algebra $\mathbb{O}$, cp. 4.26.

G of type $\mathrm{SO}(2n+1)$

G	H	$\mathrm{Cen}_G(H)^\circ$	(n_1, n_2)	$X = G/H$	Remarks
$\mathrm{SO}(2n+1)$ $n \geq 3$	$\mathrm{SO}(2n-3)$	$\mathrm{SO}(4)$	$(4n-5, 4n-1)$	$V_4(\mathbb{R}^{2n+1})$	$\pi_{2n-3} = \mathbb{Z}/2$
$\mathrm{Spin}(11)$	$\mathrm{Spin}(7)$	$\mathrm{Sp}(1)$	$(15, 19)$		$\pi_9 \neq 0$
$\mathrm{Spin}(9)$	$\mathrm{Sp}(2)$	$\mathrm{Sp}(1)$	$(11, 15)$		$\pi_5 \neq 0$
$\mathrm{Spin}(9)$	G_2	$\mathrm{SO}(2)$	$(7, 15)$	$V_2(\mathbb{O}^2)$	
$\mathrm{Spin}(7)$	$\mathrm{SU}(2)$	$\mathrm{SU}(2) \times \mathrm{SO}(3)$	$(7, 11)$	$V_3(\mathbb{R}^8)$	$\pi_5 = \mathbb{Z}/2$
$\mathrm{Spin}(7)$	${}^{\mathbb{R}}\rho_{\lambda_1} + {}^{\mathbb{R}}\rho_{2\lambda_1}$	1	$(7, 11)$		$\pi_3 = \mathbb{Z}/3$
$\mathrm{Spin}(7)$	$2 \cdot {}^{\mathbb{R}}\rho_{2\lambda_1}$	$\mathrm{SO}(2)$	$(7, 11)$		$\pi_3 = \mathbb{Z}/4$
$\mathrm{Spin}(7)$	${}^{\mathbb{R}}\rho_{4\lambda_1}$	$\mathrm{SO}(2)$	$(7, 11)$		$\pi_3 = \mathbb{Z}/10$
$\mathrm{Spin}(7)$	${}^{\mathbb{R}}\rho_{6\lambda_1}$	1	$(7, 11)$		$\pi_3 = \mathbb{Z}/28$
$\mathrm{Sp}(2)$	1	$\mathrm{Sp}(2)$	$(3, 7)$	$V_2(\mathbb{H}^2)$	

The subgroup $\mathrm{Spin}(7) \subseteq \mathrm{SO}(8)$ acts transitively on $V_3(\mathbb{R}^8)$, hence

$$\mathrm{Spin}(7)/\mathrm{SU}(2) = \mathrm{SO}(8)/\mathrm{SO}(5).$$

From the collapsing of the spectral sequence of

$$\begin{array}{ccc} \mathbb{S}^{15} = \mathrm{Spin}(9)/\mathrm{Spin}(7) & \longrightarrow & \mathrm{Spin}(11)/\mathrm{Spin}(7) \\ & & \downarrow \\ & & \mathrm{Spin}(11)/\mathrm{Spin}(9) = V_2(\mathbb{R}^{11}) \end{array}$$

one sees that

$$\begin{aligned} &\mathbf{H}^\bullet(\mathrm{Spin}(11)/\mathrm{Spin}(7); \mathbb{Z}/2) = \textstyle\bigwedge_{\mathbb{Z}/2}(x_9, x_{15}) \otimes (\mathbb{Z}/2)[x_{10}]/(x_{10}^2) \\ \neq &\mathbf{H}^\bullet(V_4(\mathbb{R}^{11}); \mathbb{Z}/2) = \textstyle\bigwedge_{\mathbb{Z}/2}(x_7, x_9) \otimes (\mathbb{Z}/2)[x_8, x_{10}]/(x_8^2, x_{10}^2). \end{aligned}$$

Similarly, the spectral sequence of

$$\begin{array}{ccc} V_2(\mathbb{R}^7) = \mathrm{Spin}(7)/\mathrm{Sp}(2) & \longrightarrow & \mathrm{Spin}(9)/\mathrm{Sp}(2) \\ & & \downarrow \\ & & \mathrm{Spin}(9)/\mathrm{Spin}(7) = \mathbb{S}^{15} \end{array}$$

collapses, and

$$\begin{aligned} &\mathbf{H}^\bullet(\mathrm{Spin}(9)/\mathrm{Sp}(2); \mathbb{Z}/2) = \textstyle\bigwedge_{\mathbb{Z}/2}(x_5, x_{15}) \otimes (\mathbb{Z}/2)[x_6]/(x_6^2) \\ \neq &\mathbf{H}^\bullet(V_4(\mathbb{R}^9); \mathbb{Z}/2) = \textstyle\bigwedge_{\mathbb{Z}/2}(x_5, x_7) \otimes (\mathbb{Z}/2)[x_6, x_8]/(x_6^2, x_8^2). \end{aligned}$$

5.3. G of type $\mathfrak{c}_n$

The homogeneous generators of the rational cohomology of $\mathrm{Sp}(n)$ have degrees

$$(3, 7, \ldots, 4n-1).$$

If $n \geq 10$, then the only types of simple groups which fit in are $\mathfrak{b}_{n-2}$ and $\mathfrak{c}_{n-2}$. However, $\mathbb{H}^n$ is not a non-trivial $\mathrm{Spin}(2n-3)$-module for $n \geq 5$, cp. 4.12, so $(\mathfrak{c}_n, \mathfrak{b}_{n-2})$ is excluded in this range, and $\mathfrak{b}_2 \cong \mathfrak{c}_2$ anyway. In low dimensions, we have also to consider the pairs $(\mathfrak{c}_9, \mathfrak{e}_7)$, $(\mathfrak{c}_6, \mathfrak{f}_4)$, and $(\mathfrak{c}_4, \mathfrak{g}_2)$.

$(\mathfrak{c}_n, \mathfrak{c}_{n-2})$ If $n \geq 4$, then $(G, H) = \mathrm{Sp}(n), \mathrm{Sp}(n-2))$, cp. 4.14.

The group $\mathrm{Sp}(1)$ has several non-trivial representation on $\mathbb{H}^3$, corresponding to the inclusions $\mathrm{Sp}(1) \subseteq \mathrm{Sp}(3)$ (i.e. ${}^{\mathbb{H}}\rho_{\lambda_1}$), $\mathrm{SU}(2) \subseteq \mathrm{Sp}(2) \subseteq \mathrm{Sp}(3)$ (i.e. $2 \cdot {}^{\mathbb{H}}\rho_{\lambda_1}$), $\mathrm{SO}(3) \subseteq \mathrm{Sp}(3)$ (i.e. ${}^{\mathbb{H}}\rho_{2\lambda_1}$), and $3 \cdot {}^{\mathbb{H}}\rho_{\lambda_1}$, ${}^{\mathbb{H}}\rho_{3\lambda_1}$, ${}^{\mathbb{H}}\rho_{3\lambda_1} + {}^{\mathbb{H}}\rho_{\lambda_1}$, ${}^{\mathbb{H}}\rho_{5\lambda_1}$, cp. 4.8.

$(\mathfrak{c}_9, \mathfrak{e}_7)$, $(\mathfrak{c}_6, \mathfrak{f}_4)$, and $(\mathfrak{c}_4, \mathfrak{g}_2)$ The compact groups E_7, F_4, and G_2 do not have non-trivial representations on $\mathbb{H}^9$, $\mathbb{H}^6$, and $\mathbb{H}^4$, respectively, thus the pairs $(\mathfrak{c}_9, \mathfrak{e}_7)$, $(\mathfrak{c}_6, \mathfrak{f}_4)$, and $(\mathfrak{c}_4, \mathfrak{g}_2)$ do not exist, cp. 4.20, 4.24, 4.26.

G of type $\mathrm{Sp}(n)$

G	H	$\mathrm{Cen}_G(H)^\circ$	(n_1, n_2)	G/H	Remarks
$\mathrm{Sp}(n)$ $n \geq 3$	$\mathrm{Sp}(n-2)$	$\mathrm{Sp}(2)$	$(4n-5, 4n-1)$	$V_2(\mathbb{H}^n)$	
$\mathrm{Sp}(3)$	$\mathrm{SO}(3)$	1	$(7, 11)$		$\pi_3 = \mathbb{Z}/8$
$\mathrm{Sp}(3)$	$2 \cdot {}^{\mathbb{H}}\rho_{\lambda_1}$	$\mathrm{Sp}(1)$	$(7, 11)$		$\pi_3 = \mathbb{Z}/2$
$\mathrm{Sp}(3)$	$3 \cdot {}^{\mathbb{H}}\rho_{\lambda_1}$	1	$(7, 11)$		$\pi_3 = \mathbb{Z}/3$
$\mathrm{Sp}(3)$	${}^{\mathbb{H}}\rho_{3\lambda_1}$	$\mathrm{Sp}(1)$	$(7, 11)$		$\pi_3 = \mathbb{Z}/10$
$\mathrm{Sp}(3)$	${}^{\mathbb{H}}\rho_{3\lambda_1} + {}^{\mathbb{H}}\rho_{\lambda_1}$	1	$(7, 11)$		$\pi_3 = \mathbb{Z}/11$
$\mathrm{Sp}(3)$	${}^{\mathbb{H}}\rho_{5\lambda_1}$	1	$(7, 11)$		$\pi_3 = \mathbb{Z}/35$
$\mathrm{Sp}(2)$	1	$\mathrm{Sp}(2)$	$(3, 7)$	$V_2(\mathbb{H}^2)$	

5.4. G of type $\mathfrak{d}_n$

The homogeneous generators of the rational cohomology of $\mathrm{Spin}(2n)$ have degrees

$$(3, 7, \ldots, 4n-5, 2n-1).$$

If $n \geq 5$, then the only types of simple groups which fit in are $\mathfrak{b}_{n-2}$ and $\mathfrak{c}_{n-2}$. However, $\mathbb{R}^{2n}$ is not a non-trivial $\mathrm{Sp}(n-2)$-module for $n \geq 4$, cp. 4.14, so $(\mathfrak{d}_n, \mathfrak{c}_{n-2})$ is excluded in this range, and $\mathfrak{b}_2 \cong \mathfrak{c}_2$. We have also to consider the pair $(\mathfrak{d}_4, \mathfrak{g}_2)$.

$(\mathfrak{d}_n, \mathfrak{b}_{n-2})$ If $n \geq 6$, then $(G, H) = (\mathrm{Spin}(2n), \mathrm{Spin}(2n-3))$, cp. 4.12.

The two representations of $\mathrm{Spin}(7)$ on $\mathbb{R}^{10}$ yield two embeddings $\mathrm{Spin}(7) \subseteq \mathrm{Spin}(10)$. Similarly, the inclusions $\mathrm{SO}(5) \subseteq \mathrm{SO}(8)$ and $\mathrm{Sp}(2) \subseteq \mathrm{SO}(8)$ yield two inclusions, cp. 4.12. The case of $\mathfrak{d}_3 \cong \mathfrak{a}_3$ was already considered above.

$(\mathfrak{d}_4, \mathfrak{g}_2)$ There is a unique representation of G_2 on $\mathbb{R}^8$, cp. 4.26.

G of type SO($2n$)

G	H	$\mathrm{Cen}_G(H)^\circ$	(n_1, n_2)	G/H	Remarks
SO($2n$) $n \geq 3$	SO($2n-3$)	SO(3)	$(2n-1, 4n-5)$	$V_3(\mathbb{R}^{2n})$	$\pi_{2n-3} = \mathbb{Z}/2$
Spin(10)	Spin(7)	SO(2)	$(9, 15)$		
Spin(8)	G_2	1	$(7, 7)$	$\mathbb{S}^7 \times \mathbb{S}^7$	
SU(4)	SU(2)	U(2)	$(5, 7)$	$V_2(\mathbb{C}^4)$	
SU(4)	SO(3)	U(1)	$(5, 7)$		$\pi_3 = \mathbb{Z}/4$
SU(4)	$\rho_{3\lambda_1}$	1	$(5, 7)$		$\pi_3 = \mathbb{Z}/10$

The spectral sequence of the bundle

$$\begin{array}{ccc} \mathbb{S}^{15} = \mathrm{Spin}(9)/\mathrm{Spin}(7) & \longrightarrow & \mathrm{Spin}(10)/\mathrm{Spin}(7) \\ & & \downarrow \\ & & \mathrm{Spin}(10)/\mathrm{Spin}(9) = \mathbb{S}^9 \end{array}$$

collapses, and thus $\mathbf{H}^\bullet(\mathrm{Spin}(10)/\mathrm{Spin}(7)) = \bigwedge_{\mathbb{Z}}(x_9, x_{15})$. In particular,

$$\begin{aligned} \mathbf{H}^\bullet(\mathrm{Spin}(10)/\mathrm{Spin}(7); \mathbb{Z}/2) &\neq \mathbf{H}^\bullet(V_3(\mathbb{R}^{10}); \mathbb{Z}/2) \\ &= \bigwedge\nolimits_{\mathbb{Z}/2}(x_7, x_9) \otimes (\mathbb{Z}/2)[x_8]/(x_8^2) \end{aligned}$$

5.5. G EXCEPTIONAL

Finally, we have the exceptional groups. Here, the only pairs are $(\mathfrak{e}_6, \mathfrak{f}_4)$, $(\mathfrak{f}_4, \mathfrak{g}_2)$, and $(\mathfrak{g}_2, 0)$.

$(\mathfrak{e}_6, \mathfrak{f}_4)$ By 4.27, there is a unique inclusion $\mathrm{F}_4 \subseteq \mathrm{E}_6$; the resulting space is Riemannian symmetric.

$(\mathfrak{f}_4, \mathfrak{g}_2)$ Again by 4.27, there is a unique inclusion $\mathrm{G}_2 \subseteq \mathrm{F}_4$.

$(\mathfrak{g}_2, 0)$

G of exceptional type

G	H	$\mathrm{Cen}_G(H)^\circ$	(n_1, n_2)	Remarks
E_6	F_4		$(9, 17)$	
F_4	G_2		$(15, 23)$	$\pi_7 = \mathbb{Z}/3$
G_2	1		$(3, 11)$	2-torsion

The $\mathbb{Z}/2$-Leray-Serre spectral sequence of the bundle

$$\begin{array}{ccc} \mathrm{SU}(2) & \longrightarrow & \mathrm{G}_2 \\ & & \downarrow \\ & & \mathrm{G}_2/\mathrm{SU}(2) = V_2(\mathbb{R}^7) \end{array}$$

collapses, and thus

$$\mathbf{H}^\bullet(\mathrm{G}_2; \mathbb{Z}/2) \cong \bigwedge\nolimits_{\mathbb{Z}/2}(x_3, x_5) \otimes (\mathbb{Z}/2)[x_6]/(x_6^2).$$

The inclusion $\mathrm{G}_2 \subseteq \mathrm{F}_4$ is analyzed in Borel [**10**]. It is proved in particular, that there is a diagram

$$\begin{array}{ccc} \mathbf{H}^\bullet(\mathrm{G}_2;\mathbb{Z}/3) & \longleftarrow & \mathbf{H}^\bullet(\mathrm{F}_4;\mathbb{Z}/3) \\ \| & & \| \\ \bigwedge_{\mathbb{Z}/3}(x_3, x_{11}) & \longleftarrow & \bigwedge_{\mathbb{Z}/3}(x_3, x_7, x_{11}, x_{15}) \otimes (\mathbb{Z}/3)[x_8]/(x_8^3), \end{array}$$

cp. *loc. cit.* 19–23. Therefore the spectral sequence of the bundle

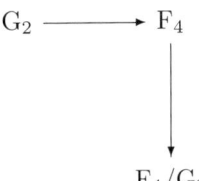

$$\begin{array}{ccc} \mathrm{G}_2 & \longrightarrow & \mathrm{F}_4 \\ & & \downarrow \\ & & \mathrm{F}_4/\mathrm{G}_2 \end{array}$$

collapses, and

$$\mathbf{H}^\bullet(\mathrm{F}_4/\mathrm{G}_2;\mathbb{Z}/3) \cong (\mathbb{Z}/3)[x_8]/(x_8^3) \otimes \bigwedge\nolimits_{\mathbb{Z}/3}(x_7, x_{15}).$$

We use this fact later.

We have classified compact 1-connected homogeneous spaces whose rational cohomology is the same as a product of odd-dimensional spheres. However, the spaces that we are interested in have the same *integral* cohomology as a product of spheres. The following theorem is the main result of this section.

THEOREM 5.6. *Let $X = G/H$ be a 1-connected homogeneous space of an effective almost simple compact Lie group G. If X has the same integral cohomology as a product of odd-dimensional spheres,*

$$\mathbf{H}^\bullet(X) \cong \bigwedge\nolimits_{\mathbb{Z}}(u, v),$$

$\deg(u) = n_1 \geq 3$ *odd,* $\deg(v) = n_2 \geq n_1$ *odd, then (G, H) is one of the following pairs.*

G	H	$\mathrm{Cen}_G(H)^\circ$	(n_1, n_2)	G/H
$\mathrm{SU}(n+1)$, $n \geq 3$	$\mathrm{SU}(n-1)$	$\mathrm{U}(2)$	$(2n-1, 2n+1)$	$V_2(\mathbb{C}^{n+1})$
$\mathrm{Sp}(n)$, $n \geq 3$	$\mathrm{Sp}(n-2)$	$\mathrm{Sp}(2)$	$(4n-5, 4n-1)$	$V_2(\mathbb{H}^n)$
E_6	F_4	1	$(9, 17)$	
$\mathrm{Spin}(10)$	$\mathrm{Spin}(7)$	$\mathrm{SO}(2)$	$(9, 15)$	
$\mathrm{Spin}(9)$	G_2	$\mathrm{SO}(2)$	$(7, 15)$	$V_2(\mathbb{O}^2)$
$\mathrm{Spin}(8)$	G_2	1	$(7, 7)$	$\mathbb{S}^7 \times \mathbb{S}^7$
$\mathrm{SU}(6)$	$\mathrm{Sp}(3)$	1	$(5, 9)$	
$\mathrm{SU}(5)$	$\mathrm{Sp}(2)$	$\mathrm{U}(1)$	$(5, 9)$	
$\mathrm{SU}(3)$	1	$\mathrm{SU}(3)$	$(3, 5)$	$V_2(\mathbb{C}^3)$
$\mathrm{Sp}(2)$	1	$\mathrm{Sp}(2)$	$(3, 7)$	$V_2(\mathbb{H}^2)$

All spaces in this list have distinct homotopy types, except for one coincidence: the subgroup $\mathrm{SU}(5) \subseteq \mathrm{SU}(6)$ *acts transitively on* $\mathrm{SU}(6)/\mathrm{Sp}(3)$*, and thus*

$$\mathrm{SU}(5)/\mathrm{Sp}(2) = \mathrm{SU}(6)/\mathrm{Sp}(3).$$

PROOF. This follows from the tables above. The space X has to be $(n_1 - 1)$-connected and torsion-free. □

5.B. Case (II): $\mathbf{H}^\bullet(X) = \mathbb{Z}[a]/(a^2) \otimes \bigwedge_\mathbb{Z}(w)$.

Now we assume that G/H has the same cohomology as a product $\mathbb{S}^{n_1} \times \mathbb{S}^{n_2}$, where $n_1 \geq 4$ is even and $n_2 > n_1$ is odd. Thus

$$\mathbf{H}^\bullet(X) \cong \mathbb{Z}[a]/(a^2) \otimes \bigwedge\nolimits_\mathbb{Z}(w),$$

where $\deg(a) = n_1$ and $\deg(w) = n_2$. If $n_1 \geq 6$, then there is an isomorphism

$$P^3_H \longleftarrow P^3_G$$

and therefore H is almost simple.

LEMMA 5.7. *If $n_1 = 4$, then G is of classical type, with $3 \leq \mathrm{rk}(G) \leq 5$. The group H is semisimple with two almost simple factors of type $\mathfrak{a}_1, \mathfrak{a}_2, \mathfrak{b}_2$, or $\mathfrak{g}_2$, respectively (we show below that a factor $\mathfrak{g}_2$ is in fact not possible).*

PROOF. The cokernel of $P^3_H \longleftarrow P^3_G$ is 1-dimensional, and $P^1_H \longleftarrow \mathcal{P}^1_G = 0$ is an isomorphism. Thus H is semisimple with two almost simple factors H_1, H_2. Next, we note that $\mathrm{rk}(\pi_7(X)) \in \{1, 2\}$. From the exact sequence

$$0 \longrightarrow \pi_7(H) \otimes \mathbb{Q} \longrightarrow \pi_7(G) \otimes \mathbb{Q} \longrightarrow \pi_7(X) \otimes \mathbb{Q} \longrightarrow 0$$

we see that

$$\mathrm{rk}(\pi_7(G)) \geq 1.$$

Moreover, $\mathrm{rk}(G) = \mathrm{rk}(H) + 1 \geq 3$. A glance at the table in 3.2 shows that G is of classical type.

We want to show that $\mathrm{rk}(G) \leq 5$. Assume otherwise. Then $\mathrm{rk}(\pi_7(G)) = 1$ (because G is of classical type), hence $\mathrm{rk}(\pi_7(H)) = 0$. By the table in 3.2, the groups H_1 and H_2 have to be of type $\mathfrak{a}_1, \mathfrak{a}_2, \mathfrak{g}_2$, or $\mathfrak{e}_6, \mathfrak{e}_7, \mathfrak{e}_8, \mathfrak{f}_4$.

We exclude the four large exceptional groups as follows. Let $\mathfrak{h}_1$ be one of these four exceptional Lie algebras, and let $\mathfrak{g}$ be any classical compact simple Lie algebra. If $\mathfrak{h}_1 \subseteq \mathfrak{g}$, then $\mathrm{rk}(\mathfrak{g}) > 2\mathrm{rk}(\mathfrak{h}_1) + 1$, cp. 4.18, 4.20, 4.22, 4.24. In our situation, $\mathrm{rk}(G) - \mathrm{rk}(H) = 1$, hence $2\mathrm{rk}(H_1) \geq \mathrm{rk}(G) - 1$ or $2\mathrm{rk}(H_2) \geq \mathrm{rk}(G) - 1$. This is not possible for these four types of groups. Therefore $\mathrm{rk}(H) \leq 4$ and thus $3 \leq \mathrm{rk}(G) \leq 5$. □

A direct computation shows that numerically, the only possibilities are $(\mathfrak{a}_3, \mathfrak{a}_1 \oplus \mathfrak{a}_1)$, $(\mathfrak{b}_3, \mathfrak{a}_1 \oplus \mathfrak{a}_1)$, $(\mathfrak{c}_3, \mathfrak{a}_1 \oplus \mathfrak{a}_1)$, $(\mathfrak{a}_4, \mathfrak{a}_1 \oplus \mathfrak{a}_2)$, $(\mathfrak{b}_4, \mathfrak{a}_1 \oplus \mathfrak{g}_2)$, $(\mathfrak{c}_4, \mathfrak{a}_1 \oplus \mathfrak{g}_2)$, $(\mathfrak{d}_4, \mathfrak{a}_1 \oplus \mathfrak{b}_2)$, $(\mathfrak{d}_4, \mathfrak{a}_1 \oplus \mathfrak{g}_2)$, $(\mathfrak{a}_5, \mathfrak{a}_2 \oplus \mathfrak{g}_2)$, and $(\mathfrak{d}_5, \mathfrak{g}_2 \oplus \mathfrak{g}_2)$.

LEMMA 5.8. *The pairs $(\mathfrak{b}_4, \mathfrak{a}_1 \oplus \mathfrak{g}_2)$, $(\mathfrak{c}_4, \mathfrak{a}_1 \oplus \mathfrak{g}_2)$, $(\mathfrak{d}_4, \mathfrak{a}_1 \oplus \mathfrak{g}_2)$, $(\mathfrak{a}_5, \mathfrak{a}_2 \oplus \mathfrak{g}_2)$, and $(\mathfrak{d}_5, \mathfrak{g}_2 \oplus \mathfrak{g}_2)$ do not exist.*

PROOF. There is only one non-trivial representation of G_2 on $\mathbb{R}^k$, $k < 14$, the standard one on $\mathbb{R}^7$, cp. 4.26. Its centralizer in $\mathrm{SO}(9)$ is $\mathrm{SO}(2)$, and this excludes $(\mathfrak{b}_4, \mathfrak{a}_1 \oplus \mathfrak{g}_2)$ and $(\mathfrak{d}_4, \mathfrak{a}_1 \oplus \mathfrak{g}_2)$. Moreover, $\mathfrak{g}_2$ is not contained in $\mathfrak{c}_4$ and $\mathfrak{a}_5$. Finally, the centralizer of G_2 in $\mathrm{SO}(10)$ is $\mathrm{SO}(3)$, and that excludes $(\mathfrak{d}_5, \mathfrak{g}_2 \oplus \mathfrak{g}_2)$. □

The following pairs remain in the case $n_1 = 4$.

$$(\mathfrak{a}_3, \mathfrak{a}_1 \oplus \mathfrak{a}_1)$$
$$(\mathfrak{a}_4, \mathfrak{a}_1 \oplus \mathfrak{a}_2)$$
$$(\mathfrak{b}_3, \mathfrak{a}_1 \oplus \mathfrak{a}_1)$$
$$(\mathfrak{c}_3, \mathfrak{a}_1 \oplus \mathfrak{a}_1)$$
$$(\mathfrak{d}_4, \mathfrak{a}_1 \oplus \mathfrak{b}_2)$$

5.9. G of type $\mathfrak{a}_n$

Here, H cannot be almost simple. The only possibilities are $(\mathfrak{a}_3, \mathfrak{a}_1 \oplus \mathfrak{a}_1)$ and $(\mathfrak{a}_4, \mathfrak{a}_1 \oplus \mathfrak{a}_2)$.

$(\mathfrak{a}_3, \mathfrak{a}_1 \oplus \mathfrak{a}_1)$ The only subgroup of type $\mathfrak{a}_1$ in SU(4) with large centralizer is the standard inclusion $\mathrm{SU}(2) \subseteq \mathrm{SU}(4)$, cp 4.8. Therefore, there is only the standard inclusion $\mathrm{SU}(2) \times \mathrm{SU}(2) \subseteq \mathrm{SU}(4)$. Note that $\mathrm{SU}(4)/\mathrm{SU}(2) \times \mathrm{SU}(2) = \mathrm{SO}(6)/\mathrm{SO}(4)$.

$(\mathfrak{a}_4, \mathfrak{a}_1 \oplus \mathfrak{a}_2)$ By 4.10, we have the standard embedding $\mathrm{SU}(3) \subseteq \mathrm{SU}(5)$, therefore we obtain the standard inclusion $\mathrm{SU}(2) \times \mathrm{SU}(3) \subseteq \mathrm{SU}(5)$ as the only possibility.

G of type $\mathrm{SU}(n)$

G	H	$\mathrm{Cen}_G(H)^\circ$	G/H	(n_1, n_2)
SU(5)	SU(3) × SU(2)	U(1)	$\widetilde{G}_3(\mathbb{C}^5)$	(4, 9)
SO(6)	SO(4)	SO(2)	$V_2(\mathbb{R}^6)$	(4, 5)

5.10. G of type $\mathfrak{b}_n$

The only possibilities are $(\mathfrak{b}_n, \mathfrak{d}_{n-1})$, for $n \geq 4$, $(\mathfrak{b}_3, \mathfrak{a}_2)$, and $(\mathfrak{b}_3, \mathfrak{a}_1 \oplus \mathfrak{a}_1)$.

$(\mathfrak{b}_n, \mathfrak{d}_{n-1})$ If $n \geq 5$, then $(G, H) = (\mathrm{SO}(2n+1), \mathrm{SO}(2n-2))$, cp. 4.16. For $n = 4$ we have the inclusions $\mathrm{SU}(4) \subseteq \mathrm{SO}(9)$ and $\mathrm{SO}(6) \subseteq \mathrm{SO}(9)$.

$(\mathfrak{b}_3, \mathfrak{a}_2)$ By 4.10, $(G, H) = (\mathrm{Spin}(7), \mathrm{SU}(3))$.

$(\mathfrak{b}_3, \mathfrak{a}_1 \oplus \mathfrak{a}_1)$ The subgroups of type $\mathfrak{a}_1$ in SO(7) with large centralizers are SO(3), with connected centralizer SO(4), and SU(2), with connected centralizer $\mathrm{SO}(3) \times \mathrm{SU}(2)$, cp. 4.8. Hence we obtain subgroups $\mathrm{SO}(3) \times \mathrm{SO}(3)$ and $\mathrm{SO}(3) \times \mathrm{SU}(2)$ in the first case. From SU(2) we obtain in addition $\mathrm{SU}(2) \cdot \mathrm{SU}(2) = \mathrm{SO}(4)$ and another copy of SO(4), obtained by mapping SU(2) diagonally into $\mathrm{SO}(3) \times \mathrm{SU}(2)$. This last group is the copy of SO(4) which is contained in $\mathrm{G}_2 = \mathrm{SO}(7) \cap \mathrm{Spin}(7) \subseteq \mathrm{SO}(8)$, and $\mathrm{Spin}(7)/\mathrm{SO}(4) = \widetilde{G}_3(\mathbb{R}^8)$.

G of type SO$(2n+1)$

G	H	$\mathrm{Cen}_G(H)^\circ$	G/H	(n_1, n_2)	Remarks
SO$(2n+1)$, $n \geq 3$	SO$(2n-2)$	SO(3)	$V_3(\mathbb{R}^{2n+1})$	$(2n-2, 4n-1)$	
Spin(9)	SU(4)	U(1)		$(6, 15)$	
Spin(7)	SU(3)	U(1)	$V_2(\mathbb{R}^8)$	$(6, 7)$	
SO(7)	SO$(3) \times$ SO(3)	1		$(4, 11)$	$\pi_2 = \mathbb{Z}/2$
SO(7)	SO$(3) \times$ SU(2)	1		$(4, 11)$	$\pi_2 = \mathbb{Z}/2$
Spin(7)	SO(4)	1	$\widetilde{G}_3(\mathbb{R}^8)$	$(4, 11)$	$\pi_2 = \mathbb{Z}/2$

The Leray-Serre spectral sequence of the bundle

$$\begin{array}{ccc} \mathbb{S}^6 = \mathrm{Spin}(7)/\mathrm{SU}(4) & \longrightarrow & \mathrm{Spin}(9)/\mathrm{SU}(4) \\ & & \big\downarrow \\ & & \mathrm{Spin}(9)/\mathrm{Spin}(7) = \mathbb{S}^{15} \end{array}$$

collapses, and thus

$$\mathbf{H}^\bullet(\mathrm{Spin}(9)/\mathrm{SU}(4)) \cong \mathbb{Z}[x_6]/(x_6^2) \otimes \bigwedge\nolimits_{\mathbb{Z}}(x_{15}).$$

The group Spin$(7) \subseteq$ SO(8) acts transitively on $V_2(\mathbb{R}^8)$; thus

$$\mathrm{Spin}(7)/\mathrm{SU}(3) = \mathrm{SO}(8)/\mathrm{SO}(6).$$

Similarly, Spin(7) acts transitively on $\widetilde{G}_3(\mathbb{R}^8)$ and

$$\mathrm{Spin}(7)/\mathrm{SO}(4) = \mathrm{SO}(8)/\mathrm{SO}(5) \times \mathrm{SO}(3).$$

5.11. G of type $\mathfrak{c}_n$

The only possibilities are $(\mathfrak{c}_n, \mathfrak{d}_{n-1})$, for $n \geq 4$, $(\mathfrak{c}_3, \mathfrak{a}_2)$, and $(\mathfrak{c}_3, \mathfrak{a}_1 \oplus \mathfrak{a}_1)$.

$(\mathfrak{c}_n, \mathfrak{d}_{n-1})$ There is no non-trivial representation of Spin$(2n-2)$ on $\mathbb{H}^n$ for $n \geq 5$ by 4.16. Therefore these pairs do not exist. There is only one non-trivial representation of SU(4) on $\mathbb{H}^4$, arising from the standard inclusion SU$(4) \subseteq$ Sp(4), cp. 4.10.

$(\mathfrak{c}_3, \mathfrak{a}_2)$ There is only one non-trivial representation of SU(3) on $\mathbb{H}^3$, arising from the standard inclusion SU$(3) \subseteq$ Sp(3), cp. 4.10.

$(\mathfrak{c}_3, \mathfrak{a}_1 \oplus \mathfrak{a}_1)$ The only representations of Sp(1) on $\mathbb{H}^3$ with large centralizers in Sp(3) are Sp(1), with connected centralizer Sp(2), and ${}^{\mathbb{H}}\rho_{3\lambda_1}$, with connected centralizer Sp(1), cp. 4.8. Thus, the possible groups are Sp$(1) \times$ Sp(1) and Sp$(1) \times {}^{\mathbb{H}}\rho_{3\lambda_1}(\mathrm{Sp}(1))$.

G of type Sp(n)

G	H	$\mathrm{Cen}_G(H)^\circ$	(n_1, n_2)	Remarks
Sp(4)	SU(4)	1	$(6, 15)$	$\pi_3 = \mathbb{Z}/2$
Sp(3)	SU(3)	1	$(6, 7)$	$\pi_3 = \mathbb{Z}/2$
Sp(3)	Sp$(1) \times$ Sp(1)	Sp(1)	$(4, 11)$	
Sp(3)	Sp$(1) \times {}^{\mathbb{H}}\rho_{3\lambda_1}(\mathrm{Sp}(1))$	1	$(4, 11)$	

The Leray-Serre spectral sequence of the bundle

$$\begin{array}{ccc} \mathbb{S}^4 = \mathrm{Sp}(2)/\mathrm{Sp}(1)\times\mathrm{Sp}(1) & \longrightarrow & \mathrm{Sp}(3)/\mathrm{Sp}(1)\times\mathrm{Sp}(1) \\ & & \downarrow \\ & & \mathrm{Sp}(3)/\mathrm{Sp}(2) = \mathbb{S}^{11} \end{array}$$

collapses, and thus

$$\mathbf{H}^\bullet(\mathrm{Sp}(3)/\mathrm{Sp}(1)\times\mathrm{Sp}(1)) \cong \mathbb{Z}[x_4]/(x_4^2)\otimes \textstyle\bigwedge_{\mathbb{Z}}(x_{11}).$$

5.12. G OF TYPE $\mathfrak{d}_n$

The only possibilities are $(\mathfrak{d}_n, \mathfrak{d}_{n-1})$, for $n \geq 4$, $(\mathfrak{d}_5, \mathfrak{a}_4)$, and $(\mathfrak{d}_4, \mathfrak{a}_1 \oplus \mathfrak{b}_2)$.

$\boxed{(\mathfrak{d}_n, \mathfrak{d}_{n-1})}$ If $n \geq 6$, then $(G, H) = (\mathrm{SO}(2n), \mathrm{SO}(2n-2))$, cp. 4.16. For SU(4) we obtain $\mathrm{SU}(4) \subseteq \mathrm{SO}(8)$ and $\mathrm{SO}(6) \subseteq \mathrm{SO}(8)$, cp. 4.10. Both inclusions become equal in Spin(8) under an automorphism of Spin(8); hence we have also only the pair (SO(8), SO(6)).

$\boxed{(\mathfrak{d}_5, \mathfrak{a}_4)}$ Then $(G, H) = (\mathrm{Spin}(10), \mathrm{SU}(5))$, cp. 4.10.

$\boxed{(\mathfrak{d}_4, \mathfrak{a}_1 \oplus \mathfrak{b}_2)}$ The non-trivial representations of Sp(2) on $\mathbb{R}^8$ are Sp(2) and SO(5), cp. 4.12. Both inclusions become equal in Spin(8) under an automorphism, hence we have the pair $(\mathrm{SO}(8), \mathrm{SO}(5)\times\mathrm{SO}(3))$.

$\boxed{G \text{ of type } \mathrm{SO}(2n)}$

G	H	$\mathrm{Cen}_G(H)^\circ$	(n_1, n_2)	G/H	Remarks
$\mathrm{SO}(2n)$ $n \geq 4$	$\mathrm{SO}(2n-2)$	$\mathrm{SO}(2)$	$(2n-2, 2n-1)$	$V_2(\mathbb{R}^{2n})$	
Spin(10)	SU(5)	U(1)	(6, 15)		(see below)
SO(8)	$\mathrm{SO}(5)\times\mathrm{SO}(3)$	1	(4, 11)	$\widetilde{G}_3(\mathbb{R}^8)$	$\pi_2 = \mathbb{Z}/2$

The subgroup $\mathrm{SO}(9) \subseteq \mathrm{SO}(10)$ acts transitively on SO(10)/SU(5); consequently,

$$\mathrm{Spin}(9)/\mathrm{SU}(4) = \mathrm{Spin}(10)/\mathrm{SU}(5).$$

Similarly,

$$\mathrm{Spin}(7)/\mathrm{SU}(3) = \mathrm{SO}(8)/\mathrm{SO}(6) = V_2(\mathbb{R}^8)$$

and

$$\mathrm{Spin}(7)/\mathrm{SO}(4) = \mathrm{SO}(8)/\mathrm{SO}(5)\times\mathrm{SO}(3) = \widetilde{G}_3(\mathbb{R}^8).$$

5.13. G EXCEPTIONAL AND H ALMOST SIMPLE

The only possibilities are $(\mathfrak{f}_4, \mathfrak{b}_3)$ and $(\mathfrak{f}_4, \mathfrak{c}_3)$. The maximal connected subgroups of maximal rank of F_4 are Spin(9), $\mathrm{SU}(3)\times\mathrm{SU}(3)/\{\pm1\}$, $\mathrm{Sp}(3)\times\mathrm{Sp}(1)/\{\pm1\}$, see Borel-De Siebenthal [**13**]. According to Dynkin [**29**] and Seitz [**89**], the maximal subgroups of strictly smaller rank are of type $\mathfrak{a}_1$, $\mathfrak{g}_2$ or $\mathfrak{a}_1 \oplus \mathfrak{g}_2$; none of these has a subgroup of type $\mathfrak{b}_3$ or $\mathfrak{c}_3$.

$\boxed{(\mathfrak{f}_4, \mathfrak{b}_3)}$ Then $H \subseteq \mathrm{Spin}(9)$. There are two conjugacy classes of groups of type Spin(7) in Spin(9); however, both groups are conjugate in F_4, as one can see by inspecting their respective fixed point sets in the Cayley plane $\mathbb{O}\mathrm{P}^2$.

$\boxed{(\mathfrak{f}_4, \mathfrak{c}_3)}$ Then $H = \mathrm{Sp}(3)$. There is one conjugacy class of subgroups of this type.

$\boxed{G \text{ of type } \mathrm{F}_4}$

G	H	$\mathrm{Cen}_G(H)^\circ$	(n_1, n_2)	Remarks
F_4	$\mathrm{Spin}(7)$	$\mathrm{SO}(2)$	$(8, 23)$	(see below)
F_4	$\mathrm{Sp}(3)$	$\mathrm{Sp}(1)$	$(8, 23)$	$\pi_5 = \mathbb{Z}/2$

The inclusion $\mathrm{G}_2 \subseteq \mathrm{Spin}(7) \subseteq \mathrm{F}_4$ in $\mathbb{Z}/3$-cohomology is according to Borel [**10**] given by the diagram

$$\begin{array}{ccccc}
\mathbf{H}^\bullet(\mathrm{G}_2; \mathbb{Z}/3) & \longleftarrow & \mathbf{H}^\bullet(\mathrm{Spin}(7); \mathbb{Z}/3) & \longleftarrow & \mathbf{H}^\bullet(\mathrm{F}_4; \mathbb{Z}/3) \\
\| & & \| & & \| \\
\bigwedge_{\mathbb{Z}/3}(x_3, x_{11}) & \longleftarrow & \bigwedge_{\mathbb{Z}/3}(x_3, x_7, x_{11}) & \longleftarrow & \bigwedge_{\mathbb{Z}/3}(x_3, x_7, x_{11}, x_{15}) \otimes (\mathbb{Z}/3)[x_8]/(x_8^3).
\end{array}$$

Therefore the $\mathbb{Z}/3$-Leray-Serre spectral sequence of the bundle

$$\begin{array}{ccc}
\mathrm{Spin}(7) & \longrightarrow & \mathrm{F}_4 \\
 & & \downarrow \\
 & & \mathrm{F}_4/\mathrm{Spin}(7)
\end{array}$$

collapses, and

$$\mathbf{H}^\bullet(\mathrm{F}_4/\mathrm{Spin}(7); \mathbb{Z}/3) \cong \bigwedge\nolimits_{\mathbb{Z}/3}(x_{15}) \otimes (\mathbb{Z}/3)[x_8]/(x_8^3).$$

The following theorem summarizes the discussion of case (II).

THEOREM 5.14. *Let $X = G/H$ be a 1-connected homogeneous space of an effective almost simple compact Lie group G. If X has the same integral cohomology as a product of an even- and an odd-dimensional sphere,*

$$\mathbf{H}^\bullet(X) \cong \mathbb{Z}[a] \otimes \bigwedge\nolimits_{\mathbb{Z}}(w),$$

$\deg(a) = n_1 \geq 4$, $\deg(w) = n_2 > n_1$, *then (G, H) is one of the following pairs.*

G	H	$\mathrm{Cen}_G(H)^\circ$	(n_1, n_2)	G/H
$\mathrm{SO}(2n)$ $n \geq 3$	$\mathrm{SO}(2n-2)$	$\mathrm{SO}(2)$	$(2n-2, 2n-1)$	$V_2(\mathbb{R}^{2n})$
$\mathrm{Spin}(10)$	$\mathrm{SU}(5)$	$\mathrm{U}(1)$	$(6, 15)$	
$\mathrm{Spin}(9)$	$\mathrm{SU}(4)$	$\mathrm{U}(1)$	$(6, 15)$	
$\mathrm{Spin}(7)$	$\mathrm{SU}(3)$	$\mathrm{U}(1)$	$(6, 7)$	$V_2(\mathbb{R}^8)$
$\mathrm{Sp}(3)$	$\mathrm{Sp}(1) \times \mathrm{Sp}(1)$	$\mathrm{Sp}(1)$	$(4, 11)$	
$\mathrm{Sp}(3)$	$\mathrm{Sp}(1) \times {}^{\mathbb{H}}\rho_{3\lambda_1}(\mathrm{Sp}(1))$	$\mathrm{Sp}(1)$	$(4, 11)$	
$\mathrm{SU}(5)$	$\mathrm{SU}(3) \times \mathrm{SU}(2)$	$\mathrm{U}(1)$	$(4, 9)$	$\widetilde{G}_3(\mathbb{C}^5)$

There are two coincidences in this table;

$$\mathrm{Spin}(9)/\mathrm{SU}(4) = \mathrm{Spin}(10)/\mathrm{SU}(5)$$

and

$$\mathrm{Spin}(7)/\mathrm{SU}(3) = \mathrm{SO}(8)/\mathrm{SO}(6) = V_2(\mathbb{R}^8).$$

PROOF. This follows from the discussion of the various groups. All other spaces are either not $(n_1 - 1)$-connected, or have torsion. □

CHAPTER 6

The case when G is semisimple

We continue the classification of homogeneous spaces $X = G/H$ which have the same cohomology as a product of spheres. In this chapter we assume that G is semisimple with two almost simple factors (by 3.14, this is the remaining case). More precisely, our assumptions are as follows.

(i) G/H is compact and 1-connected.
(ii) G is a product of two compact almost simple Lie groups K_1, K_2.
(iii) the action of G on X irreducible and almost effective.
(iv) G/H has the same integral cohomology as a product $\mathbb{S}^{n_1} \times \mathbb{S}^{n_2}$, where $3 \leq n_1 \leq n_2$, and n_2 is odd.

Note that H has to be connected. Put

$$H_i = (H \cap K_i)^\circ.$$

Then H_1, H_2 are normal subgroups in H, and thus there exists a compact connected normal subgroup $H_0 \subseteq H$ such that H is an almost direct product

$$H = (H_1 \times H_2) \cdot H_0.$$

If $H_0 = 1$, then

$$G/H = K_1/H_1 \times K_2/H_2$$

is split. We call this the **split case**.

If $H_0 \neq 1$, then we consider again the two cases

Case (I): n_1 is odd, $\mathbf{H}^\bullet(G/H) \cong \bigwedge_{\mathbb{Z}}(u, v)$, with $\deg(u) = n_1$ and $\deg(v) = n_2$. In this situation Theorem 3.7 applies.

Case (II): n_1 is even, $\mathbf{H}^\bullet(G/H) \cong \mathbb{Z}[a]/(a^2) \otimes \bigwedge_{\mathbb{Z}}(u)$, with $\deg(a) = n_1$ and $\deg(u) = n_2$. In this situation Theorem 3.11 applies.

These three cases are considered in the next sections. Each section ends with a complete list of all possible pairs (G, H).

6.A. The split case

Here,

$$X = K_1/H_1 \times K_2/H_2$$

is a product. From the Künneth theorem we see that K_1/H_1 and K_2/H_2 have to be 1-connected cohomology spheres, say, of dimensions n_1, n_2. Thus we have to determine all pairs (K, H) with the following properties.

(i) K is a compact almost simple Lie group.
(ii) K/H is a 1-connected cohomology m-sphere.
(iii) the action of K on K/H is effective.

If m is even, then K/H has Euler characteristic 2, and a well-known result of Borel-De Siebenthal [**13**] and Borel [**8**] says that K/H is actually a sphere, and that (K, H) is one of the following pairs.

Homogeneous spaces of Euler characteristic 2

K	H	$\mathrm{Cen}_K(H)^\circ$	m	K/H
$\mathrm{SO}(2n+1)$ $n \geq 1$	$\mathrm{SO}(2n)$	1	$2n$	$\mathbb{S}^{2n}$
G_2	$\mathrm{SU}(3)$	1	6	$\mathbb{S}^6$

We need a similar result for 1-connected homogeneous spaces K/H with

$$\mathbf{H}^\bullet(K/H; \mathbb{Z}) = \bigwedge\nolimits_{\mathbb{Z}}(w),$$

where $\deg(w) = m \geq 3$ is odd. This has been obtained by various authors, cp. e.g. Bredon [**15**], Onishchik [**80**] Ch. 5 §18 Table 10. For the sake of completeness, we give a proof. The method is very similar to case (I) in the previous chapter. The group G has to be almost simple by a similar reasoning as in 3.14, and H has to be trivial or almost simple, because of the isomorphism $P^1_H \xleftarrow{\cong} P^1_G = 0$ and the surjection $P^3_H \longleftarrow P^3_G$, cp. 3.7.

6.1. K of type $\mathfrak{a}_n$

The only possibilities are $(\mathfrak{a}_n, \mathfrak{a}_{n-1})$ and $(\mathfrak{a}_3, \mathfrak{b}_2)$.

$(\mathfrak{a}_n, \mathfrak{a}_{n-1})$ If $n \geq 3$, then $(K, H) = (\mathrm{SU}(n+1), \mathrm{SU}(n))$ by 4.10. For $n = 2$, there is also the pair $(\mathrm{SU}(3), \mathrm{SO}(3))$.

$(\mathfrak{a}_3, \mathfrak{b}_2)$ By 4.12, the only possibility is $(\mathrm{SO}(6), \mathrm{SO}(5))$.

K of type $\mathrm{SU}(n+1)$

K	H	$\mathrm{Cen}_K(H)^\circ$	m	K/H	Remarks
$\mathrm{SU}(n+1)$ $n \geq 2$	$\mathrm{SU}(n)$	$\mathrm{U}(1)$	$2n+1$	$\mathbb{S}^{2n+1}$	
$\mathrm{SO}(6)$	$\mathrm{SO}(5)$	1	5	$\mathbb{S}^5$	
$\mathrm{SU}(3)$	$\mathrm{SO}(3)$	1	5		$\pi_3 = \mathbb{Z}/4$
$\mathrm{SU}(2)$	1	$\mathrm{SU}(2)$	3	$\mathbb{S}^3$	

6.2. K of type $\mathfrak{b}_n$

The only possibilities are $(\mathfrak{b}_n, \mathfrak{b}_{n-1})$, $(\mathfrak{b}_n, \mathfrak{c}_{n-1})$ and $(\mathfrak{b}_3, \mathfrak{g}_2)$.

$(\mathfrak{b}_n, \mathfrak{c}_{n-1})$ If $n \geq 4$, then there is no non-trivial representation of $\mathrm{Sp}(n-1)$ on $\mathbb{R}^{2n+1}$ by 4.14, hence these pairs do not exist.

$(\mathfrak{b}_n, \mathfrak{b}_{n-1})$ If $n \geq 5$, then $(K, H) = (\mathrm{SO}(2n+1), \mathrm{SO}(2n-1))$ by 4.12. For $n = 2, 3, 4$, there are also the pairs $(\mathrm{Spin}(9), \mathrm{Spin}(7))$, $(\mathrm{Spin}(7), \mathrm{Sp}(2))$, $(\mathrm{Sp}(2), \mathrm{Sp}(1))$, and $(\mathrm{Sp}(2), {}^{\mathbb{H}}\rho_{3\lambda_1}(\mathrm{Sp}(1)))$. However, $\mathrm{Spin}(7)$ contains the center of $\mathrm{Sp}(2)$, and thus $\mathrm{Spin}(7)/\mathrm{Sp}(2) = \mathrm{SO}(7)/\mathrm{SO}(5)$.

$(\mathfrak{b}_3, \mathfrak{g}_2)$ By 4.26, $(K, H) = (\mathrm{Spin}(7), \mathrm{G}_2)$.

K of type $\mathrm{SO}(2n+1)$

K	H	$\mathrm{Cen}_K(H)^\circ$	m	K/H	Remarks
$\mathrm{SO}(2n+1)$ $n \geq 2$	$\mathrm{SO}(2n-1)$	$\mathrm{SO}(2)$	$4n-1$	$V_2(\mathbb{R}^{2n+1})$	$\pi_{2n-1} = \mathbb{Z}/2$
$\mathrm{Spin}(9)$	$\mathrm{Spin}(7)$	1	15	$\mathbb{S}^{15}$	
$\mathrm{Spin}(7)$	G_2	1	7	$\mathbb{S}^7$	
$\mathrm{Sp}(2)$	$\mathrm{Sp}(1)$	$\mathrm{Sp}(1)$	7	$\mathbb{S}^7$	
$\mathrm{Sp}(2)$	${}^{\mathbb{H}}\rho_{3\lambda_1}$	1	7		$\pi_3 = \mathbb{Z}/10$

6.3. K of type $\mathfrak{c}_n$

The only possibilities are $(\mathfrak{c}_n, \mathfrak{c}_{n-1})$, $(\mathfrak{c}_n, \mathfrak{b}_{n-1})$ and $(\mathfrak{c}_3, \mathfrak{g}_2)$.

$(\mathfrak{c}_n, \mathfrak{b}_{n-1})$ If $n \geq 4$, then there is no non-trivial representation of $\mathrm{Spin}(2n-1)$ on $\mathbb{H}^n$ by 4.12, hence these pairs do not exist.

$(\mathfrak{c}_n, \mathfrak{c}_{n-1})$ If $n \geq 3$, then $(K, H) = (\mathrm{Sp}(n), \mathrm{Sp}(n-1))$ by 4.14. For $n = 2$, there are also the pairs $(\mathrm{SO}(5), \mathrm{SO}(3))$ and $(\mathrm{Sp}(2), {}^{\mathbb{H}}\rho_{3\lambda_1}(\mathrm{Sp}(1)))$.

$(\mathfrak{c}_3, \mathfrak{g}_2)$ There is no non-trivial representation of G_2 on $\mathbb{H}^3$ by 4.26.

K of type $\mathrm{Sp}(n)$

K	H	$\mathrm{Cen}_K(H)^\circ$	m	K/H	Remarks
$\mathrm{Sp}(n)$ $n \geq 2$	$\mathrm{Sp}(n-1)$	$\mathrm{Sp}(1)$	$4n-1$	$\mathbb{S}^{4n-1}$	
$\mathrm{SO}(5)$	$\mathrm{SO}(3)$	$\mathrm{SO}(2)$	7	$V_2(\mathbb{R}^5)$	$\pi_3 = \mathbb{Z}/2$
$\mathrm{Sp}(2)$	${}^{\mathbb{H}}\rho_{3\lambda_1}$	1	7		$\pi_3 = \mathbb{Z}/10$
$\mathrm{Sp}(1)$	1	$\mathrm{Sp}(1)$	3	$\mathbb{S}^3$	

6.4. K of type $\mathfrak{d}_n$

The only possibilities are $(\mathfrak{d}_n, \mathfrak{b}_{n-1})$ and $(\mathfrak{d}_n, \mathfrak{c}_{n-1})$.

$(\mathfrak{d}_n, \mathfrak{b}_{n-1})$ If $n \geq 5$, then $(K, H) = (\mathrm{SO}(2n), \mathrm{SO}(2n-1))$ by 4.12. For $n = 4$, there are is also the pair $(\mathrm{Spin}(8), \mathrm{Spin}(7))$. However, the action of $\mathrm{Spin}(8)$ on $\mathrm{Spin}(8)/\mathrm{Spin}(7)$ is not effective, and thus $\mathrm{Spin}(8)/\mathrm{Spin}(7) = \mathrm{SO}(8)/\mathrm{SO}(7)$.

$(\mathfrak{d}_n, \mathfrak{c}_{n-1})$ If $n \geq 4$, then $\mathbb{R}^{2n}$ is not a non-trivial $\mathrm{Sp}(n-1)$-module by 4.14, hence these pairs do not exist.

K of type $\mathrm{SO}(2n)$

K	H	$\mathrm{Cen}_K(H)^\circ$	m	K/H
$\mathrm{SO}(2n)$ $n \geq 3$	$\mathrm{SO}(2n-1)$	1	$2n-1$	$\mathbb{S}^{2n-1}$

6.5. K of exceptional type

The only possibility is $(\mathfrak{g}_2, \mathfrak{a}_1)$. The subgroups of type $\mathfrak{a}_1$ in G_2 were determined in 4.27.

K of exceptional type

K	H	$\mathrm{Cen}_K(H)^\circ$	m	K/H	Remarks
G_2	$\mathrm{SU}(2)$	$\mathrm{SU}(2)$	11	$V_2(\mathbb{R}^7)$	$\pi_5 = \mathbb{Z}/2$
G_2	$2 \cdot {}^{\mathbb{R}}\rho_{2\lambda_1}$	1	11		$\pi_3 = \mathbb{Z}/4$
G_2	${}^{\mathbb{R}}\rho_{\lambda_1} + {}^{\mathbb{R}}\rho_{2\lambda_1}$	$\mathrm{SU}(2)$	11		$\pi_3 = \mathbb{Z}/3$
G_2	${}^{\mathbb{R}}\rho_{6\lambda_1}$	1	11		$\pi_3 = \mathbb{Z}/28$

Later we will need the $\mathbb{Z}/2$-cohomology of G_2/H, where $H \cong \mathrm{SU}(2)$. There are two such embeddings. In both cases, the map $\mathbf{H}_3(\mathrm{SU}(2)) \longrightarrow \mathbf{H}_3(\mathrm{G}_2))$ is multiplication by an odd number (1 or 3, respectively). Thus we have an isomorphism $\mathbf{H}^3(\mathrm{SU}(2);\mathbb{Z}/2) \longleftarrow \mathbf{H}^3(\mathrm{G}_2;\mathbb{Z}/2)$. It follows that the $\mathbb{Z}/2$-Lerray-Serre spectral sequence of the bundle

$$\begin{array}{ccc} \mathrm{SU}(2) \cong H & \longrightarrow & \mathrm{G}_2 \\ & & \big\downarrow \\ & & \mathrm{G}_2/H \end{array}$$

collapses. For $H = \mathrm{SU}(2)$, the base is the Stiefel manifold $V_2(\mathbb{R}^7)$. Thus

$$\mathbf{H}^\bullet(\mathrm{G}_2;\mathbb{Z}/2) \cong \bigwedge\nolimits_{\mathbb{Z}/2}(x_3, x_5) \otimes (\mathbb{Z}/2)[x_6]/(x_6^2).$$

It follows that in both cases

$$\mathbf{H}^\bullet(\mathrm{G}_2/H;\mathbb{Z}/2) \cong \bigwedge\nolimits_{\mathbb{Z}/2}(x_5) \otimes (\mathbb{Z}/2)[x_6]/(x_6^2).$$

THEOREM 6.6. *Let K/H be a 1-connected homogeneous space of a compact connected simple Lie group K. Assume that the action of K is effective and irreducible. Assume that K/H has the same integral cohomology as an m-sphere, $m \geq 2$. Then K/H is a (standard) sphere, and (K, H) is one of the following pairs.*

Homogeneous spheres

K	H	$\mathrm{Cen}_K(H)^\circ$	K/H
$\mathrm{SO}(2n+1)$, $n \geq 1$	$\mathrm{SO}(2n)$	1	$\mathbb{S}^{2n}$
G_2	$\mathrm{SU}(3)$	1	$\mathbb{S}^6$
$\mathrm{SU}(n+1)$, $n \geq 2$	$\mathrm{SU}(n)$	$\mathrm{U}(1)$	$\mathbb{S}^{2n+1}$
$\mathrm{Sp}(n)$, $n \geq 2$	$\mathrm{Sp}(n-1)$	$\mathrm{Sp}(1)$	$\mathbb{S}^{4n-1}$
$\mathrm{SO}(2n)$, $n \geq 3$	$\mathrm{SO}(2n-1)$	1	$\mathbb{S}^{2n-1}$
$\mathrm{Spin}(9)$	$\mathrm{Spin}(7)$	1	$\mathbb{S}^{15}$
$\mathrm{Spin}(7)$	G_2	1	$\mathbb{S}^7$
$\mathrm{SU}(2)$	1	$\mathrm{SU}(2)$	$\mathbb{S}^3$

COROLLARY 6.7. *Let G/H be a 1-connected homogeneous space of a compact connected Lie group G. Assume that $G = K_1 \times K_2$ is a product of two compact*

almost simple Lie groups. Assume moreover that the action is irreducible and split, i.e. that $H = (K_1 \cap H) \times (K_2 \cap H)$. *If* G/H *has the same integral cohomology as a product of spheres* $\mathbb{S}^{n_1} \times \mathbb{S}^{n_2}$, *where* $n_1 \geq 3$ *and* $n_2 \geq n_1$ *is odd, then the factors* K_i/H_i *are spheres, and the possibilities of the pairs* (K_1, H_1), (K_2, H_2) *are given by the table above.* □

We note also the following.

THEOREM 6.8. *Let* G/H *be a 1-connected homogeneous space of a compact connected Lie group* G. *Assume that the action is effective and irreducible, and that* G/H *has the same rational cohomology as* $\mathbb{S}^{2m-3}$, *where* $m \geq 3$. *Assume that* G/H *is* $(m-2)$*-connected, and that* $\pi_{m-1}(G/H) \cong \mathbb{Z}/2$. *Then* $G/H \cong V_2(\mathbb{R}^m)$ *is a Stiefel manifold, and there are only the following possibilities.*

G	H	$\mathrm{Cen}_G(H)^\circ$	$X = G/H$
$\mathrm{SO}(2n+1)$	$\mathrm{SO}(2n-1)$	$\mathrm{SO}(2)$	$V_2(\mathbb{R}^{2n+1})$
$n \geq 2$			
G_2	$\mathrm{SU}(2)$	$\mathrm{U}(1)$	$V_2(\mathbb{R}^7)$

The following will also be useful.

COROLLARY 6.9. *Let* K/H *be a 1-connected homogeneous space of a compact connected Lie group* K. *Assume that the action is effective and irreducible, and that* K/H *has the same rational cohomology as an odd-dimensional sphere* $\mathbb{S}^m$ *where* $m \geq 3$. *Assume in addition that* $\mathrm{Cen}_K(H)$ *contains a subgroup of type* $\mathfrak{a}_1$. *Then* (K, H) *is one of the following pairs.*

K	H
$\mathrm{Sp}(n)$	$\mathrm{Sp}(n-1)$
$n \geq 2$	
G_2	$\mathrm{SU}(2)$
G_2	${}^{\mathbb{R}}\rho_{\lambda_1} + {}^{\mathbb{R}}\rho_{2\lambda_1}$
$\mathrm{Sp}(1)$	1

6.B. The non-split case (I): $\mathbf{H}^\bullet(X) = \bigwedge_{\mathbb{Z}}(u,v)$.

The setting is as described at the beginning of this chapter. Thus $G = K_1 \times K_2$ and $H = (H_1 \times H_2) \cdot H_0$. We assume that n_1 is odd, and that $H_0 \neq 1$. We first prove some general results. On the level of Lie algebras we have a direct sum of ideals $\mathfrak{h} = \mathfrak{h}_1 \oplus \mathfrak{h}_2 \oplus \mathfrak{h}_0$. Let pr_i denote the projection $\mathfrak{g} = \mathfrak{k}_1 \oplus \mathfrak{k}_2 \longrightarrow \mathfrak{k}_i$. The restriction of pr_i to $\mathfrak{h}_0 \oplus \mathfrak{h}_i$ is a monomorphism, for $i = 1, 2$. Since $\mathfrak{h}_0$ and $\mathfrak{h}_i$ commute, $[\mathfrak{h}_0, \mathfrak{h}_i] = 0$, the same is true for their images under pr_i. It follows that on the group level the restriction of pr_i to H_0 has finite kernel, and that $\mathrm{pr}_i(H_0)$ is centralized by H_i,

$$\mathrm{pr}_i(H_0) \subseteq \mathrm{Cen}_{K_i}(H_i).$$

Put $k_i = \mathrm{rk}(K_i)$ and $h_i = \mathrm{rk}(H_i)$. Since $\mathrm{pr}_i(\mathfrak{h}_0) \oplus \mathfrak{h}_i \subseteq \mathfrak{k}_i$, we have

$$k_i - h_i - h_0 \geq 0$$

for $i = 1, 2$. Note that $\mathrm{pr}_i(H_0) \neq K_i$, since otherwise the action would be reducible. In particular,

$$k_1, k_2 \geq 2$$

(otherwise pr_i would be surjective). Note also that $h_0 \geq 1$, because we assume that $H_0 \neq 1$. Thus

$$\begin{aligned} 2 &= \mathrm{rk}(G) - \mathrm{rk}(H) \\ &= k_1 - h_1 + k_2 - h_2 - h_0. \end{aligned}$$

If we subtract h_0, the result is still non-negative, hence $1 \leq h_0 \leq 2$. From the exact sequences

$$0 \longleftarrow P_H^1 \longleftarrow P_G^1 \longleftarrow 0 \quad \text{and} \quad 0 \longleftarrow P_H^3 \longleftarrow P_G^3 \longleftarrow \mathbf{H}^3(G/H;\mathbb{Q}) \longleftarrow 0$$

we see that H is semisimple with at most two almost simple factors. Therefore we may assume that $H_2 = 1$ (and thus $h_2 = 0$). The equation $2 = k_1 - h_1 - h_0 + k_2$ shows that

$$k_2 = 2$$

and

$$k_1 = h_1 + h_0.$$

Note also that $P_{H_1} \longleftarrow P_{K_1}$ is onto, because $P_H \longleftarrow P_G$ is onto.

LEMMA 6.10. *If $h_0 = 2$, then $H = H_0$ is of type $\mathfrak{a}_1 + \mathfrak{a}_1$ and $K_1, K_2 \in \{\mathrm{G}_2, \mathrm{Sp}(2)\}$. If $K_1 = K_2$, then H_0 is not conjugate to a subgroup contained in the diagonal. In particular, $G \neq \mathrm{Sp}(2) \times \mathrm{Sp}(2)$.*

PROOF. Since $\mathrm{pr}_2(H_0) \subseteq K_2$ the group K_2 has a semisimple subgroup of rank 2. The only possibilities for $(K_2, \mathrm{pr}_2(H_0))$ are thus

$$(\mathrm{G}_2, \mathrm{SO}(4)), (\mathrm{G}_2, \mathrm{SU}(3)), (\mathrm{Sp}(2), \mathrm{Sp}(1) \times \mathrm{Sp}(1)).$$

Note that H_1 has to centralize $\mathrm{pr}_1(H_0)$, and that (K_1, H_1) is one of the pairs determined in 6.9. This shows that $H_1 = 1$. Since $\mathrm{pr}_1(H_0) \neq K_1$, we obtain the possibilities $\mathrm{Sp}(2)$ and G_2 for K_1.

If H_0 is contained in the diagonal, then $\pi_3(G/H_0) \cong \mathbb{Z}$, and thus $\mathbf{H}^\bullet(G/H;\mathbb{Q})$ contains an element of degree 3, and thus the structure of the cohomology ring is not the right one (we need $\mathbf{H}^\bullet(G/H;\mathbb{Q}) \cong \bigwedge_{\mathbb{Q}}(P_G/P_G^3)$).

If $H_0 = \mathrm{SU}(3)$, then $K_1 = K_2 = \mathrm{G}_2$. Thus H_0 is conjugate to the copy of $\mathrm{SU}(3)$ which is contained in the diagonal. □

6.11. THE CASE $\mathrm{rk}(H_0) = 2$

The following possibilities remain: (1) $G = \mathrm{Sp}(2) \times \mathrm{G}_2$, and H_0 is contained in $\mathrm{Sp}(1) \times \mathrm{Sp}(1) \times \mathrm{SO}(4)$. All such groups which are not contained in one factor are conjugate. The map $\pi_3(H) \longrightarrow \pi_3(G)$ is given by the matrix $\begin{pmatrix} 1 & 3 \\ 1 & 1 \end{pmatrix}$, and thus $\pi_3(G/H) = \mathbb{Z}/2$.

(2) $G = \mathrm{G}_2 \times \mathrm{G}_2$ and $H_0 \subseteq \mathrm{SO}(4) \times \mathrm{SO}(4)$ is a subgroup which is not contained in one factor. There are two such conjugacy classes, the diagonal subgroup and the anti-diagonal subgroup. In this case the map $\pi_3(H) \longrightarrow \pi_3(G)$ is given by the matrix $\begin{pmatrix} 1 & 3 \\ 3 & 1 \end{pmatrix}$, and thus $\pi_3(G/H) = \mathbb{Z}/8$.

H_0 of type $\mathfrak{a}_1 + \mathfrak{a}_1$

G	(n_1, n_2)	Remarks
$\mathrm{G}_2 \times \mathrm{Sp}(2)$	$(7, 11)$	$\pi_3 = \mathbb{Z}/2$
$\mathrm{G}_2 \times \mathrm{G}_2$	$(11, 11)$	$\pi_3 = \mathbb{Z}/8$

LEMMA 6.12. *If $h_0 = 1$, then (K_1, H_1) is one of the pairs $(\mathfrak{c}_n, \mathfrak{c}_{n-1})$, $n \geq 2$, or $(\mathfrak{g}_2, \mathfrak{a}_1)$. The group H_0 is of type $\mathfrak{a}_1$, and K_2 is of type $\mathfrak{a}_2$, $\mathfrak{c}_2$, or $\mathfrak{g}_2$.*

PROOF. The group K_2 has rank 2, hence K_2 is one of the groups SU(3), Sp(2), G$_2$, and H_0 is of type $\mathfrak{a}_1$. Moreover, $k_1 - h_1 = 1$, and $P_{H_1} \longleftarrow P_{K_1}$ is a surjection. Thus (K_1, H_1) is one of the pairs determined in 6.9. The only possibilities are the pairs $(\mathfrak{c}_n, \mathfrak{c}_{n-1})$, $n \geq 2$, and $(\mathfrak{g}_2, \mathfrak{a}_1)$. □

Before we consider the various possibilities, we make some more general observations. It will turn out that in all cases $\mathrm{pr}_1(H_0) \cong \mathrm{SU}(2)$. Therefore we can write the elements of H_0 as pairs $(h, \phi(h))$, where $\phi : \mathrm{SU}(2) \longrightarrow \mathrm{pr}_2(H_0) \subseteq K_2$ is some fixed non-trivial homomorphism. Consider the fibre bundle which arises from the inclusions

$$H \subseteq K_1 \times \phi(H_0) \subseteq G$$

by taking quotients. Note that

$$(K_1 \times \phi(H_0))/H \cong K_1/(H_1 \cdot \ker(\phi))$$

(because the two spaces have the same dimension) and

$$G/(K_1 \times \phi(H_0)) \cong K_2/\phi(H_0).$$

Hence we obtain a fibre bundle

$$\begin{array}{ccc} K_1/(H_1 \cdot \ker(\phi)) & \longrightarrow & G/H \\ & & \downarrow \\ & & K_2/\phi(H_0), \end{array}$$

with structure group $K_1 \times \phi(H_0)$.

6.13. $h_0 = 1$ AND K_1 OF TYPE $\mathfrak{c}_n$

Then $K_1 = \mathrm{Sp}(n)$, $H_1 = \mathrm{Sp}(n-1)$, $\mathrm{pr}_1(H_0) = \mathrm{Sp}(1)$, with the standard inclusion $\mathrm{Sp}(n-1) \times \mathrm{Sp}(1) \subseteq \mathrm{Sp}(n)$, and $n \geq 2$. Thus $H_0 = \mathrm{Sp}(1)$, and the inclusion $H_0 \subseteq K_1 \times K_2$ is given by $h \longmapsto (h, \phi(h))$, where $\phi : \mathrm{Sp}(1) \longrightarrow K_2$ is a non-trivial representation.

$K_2 = \mathrm{SU}(3)$ (1) If $\phi = \rho_{\lambda_1}$ is the standard inclusion $\mathrm{SU}(2) \subseteq \mathrm{SU}(3)$, then we obtain a fibre bundle

$$\begin{array}{ccc} \mathbb{S}^{4n-1} & \longrightarrow & G/H \\ & & \downarrow \\ & & \mathbb{S}^5. \end{array}$$

The structure group of this sphere bundle is $\mathrm{Sp}(n) \times \mathrm{Sp}(1) \subseteq \mathrm{SO}(4n)$, hence it is given by a classifying map

$$\mathbb{S}^5 \longrightarrow \mathrm{B}(\mathrm{Sp}(n) \times \mathrm{Sp}(1)) \longrightarrow \mathrm{BSO}(4n).$$

Since $\pi_5(\mathrm{BSO}(4n)) \cong \pi_4(\mathrm{SO}(4n)) \cong \pi_4(\mathrm{SO}) = 0$ for $n \geq 2$, the map is homotopic to a constant map and the bundle is trivial,

$$G/H \cong \mathbb{S}^{4n-1} \times \mathbb{S}^5.$$

(2) For $\phi = {}^{\mathbb{R}}\rho_{2\lambda_1}$ we obtain an $\mathbb{RP}^{4n-1}$-bundle over $\mathrm{SU}(3)/\mathrm{SO}(3)$. Note that the higher homotopy groups of the fibre are the same as those of $\mathbb{S}^{4n-1}$. This implies that $\pi_3(G/H) \cong \pi_3(\mathrm{SU}(3)/\mathrm{SO}(3)) = \mathbb{Z}/4$.

$\boxed{K_2 = \mathrm{Sp}(2)}$ For $\phi : \mathrm{SU}(2) \longrightarrow \mathrm{Sp}(2)$ we have the three possibilities (1) $\phi = {}^{\mathbb{H}}\rho_{\lambda_1}$, (2) $\phi = 2 \cdot {}^{\mathbb{H}}\rho_{\lambda_1}$ and (3) $\phi = {}^{\mathbb{H}}\rho_{3\lambda_1}$. In all three cases the fibre of our bundle is $\mathbb{S}^{4n-1}$, hence $\pi_3(G/H) \cong \pi_3(\mathrm{Sp}(2)/\phi(\mathrm{Sp}(1)) = \mathbb{Z}/j_\phi$.

In case (1), the base of the bundle is $\mathbb{S}^7$. Similarly as above, $\pi_7(\mathrm{BSO}(4n)) \cong \pi_6(\mathrm{SO}(4n)) \cong \pi_6(\mathrm{SO}) = 0$ for $n \geq 2$, hence

$$G/H = \mathbb{S}^{4n-1} \times \mathbb{S}^7.$$

$\boxed{K_2 = \mathrm{G}_2}$ We have the three possibilities (1) $\phi = {}^{\mathbb{R}}\rho_{\lambda_1}$, (2) $\phi = {}^{\mathbb{R}}\rho_{\lambda_1} + {}^{\mathbb{R}}\rho_{2\lambda_1}$, (3) $\phi = 2 \cdot {}^{\mathbb{R}}\rho_{2\lambda_1}$, and (4) $\phi = {}^{\mathbb{R}}\rho_{6\lambda_1}$. The fibre of the bundle is $\mathbb{S}^{4n-1}$ in case (1) and (2), and $\mathbb{RP}^{4n-1}$ in case (3) and (4), hence $\pi_3(G/H) \cong \mathbb{Z}/j_\phi$.

$\boxed{K_1 \text{ of type } \mathrm{Sp}(n)}$

K_2	ϕ	(n_1, n_2)	G/H	Remarks
$\mathrm{SU}(3)$	ρ_{λ_1}	$(5, 4n-1)$	$\mathbb{S}^5 \times \mathbb{S}^{4n-1}$	
$\mathrm{SU}(3)$	$\rho_{2\lambda_1}$	$(5, 4n-1)$		$\pi_3 = \mathbb{Z}/4$
$\mathrm{Sp}(2)$	${}^{\mathbb{H}}\rho_{\lambda_1}$	$(7, 4n-1)$	$\mathbb{S}^7 \times \mathbb{S}^{4n-1}$	
$\mathrm{SO}(5)$	${}^{\mathbb{R}}\rho_{2\lambda_1}$	$(7, 4n-1)$		$\pi_3 = \mathbb{Z}/2$
$\mathrm{Sp}(2)$	${}^{\mathbb{H}}\rho_{3\lambda_1}$	$(7, 4n-1)$		$\pi_3 = \mathbb{Z}/10$
G_2	${}^{\mathbb{R}}\rho_{\lambda_1}$	$(11, 4n-1)$		$\pi_5 = \mathbb{Z}/2$
G_2	${}^{\mathbb{R}}\rho_{\lambda_1} + {}^{\mathbb{R}}\rho_{2\lambda_1}$	$(11, 4n-1)$		$\pi_3 = \mathbb{Z}/3$
G_2	$2 \cdot {}^{\mathbb{R}}\rho_{2\lambda_1}$	$(11, 4n-1)$		$\pi_3 = \mathbb{Z}/4$
G_2	${}^{\mathbb{R}}\rho_{6\lambda_1}$	$(11, 4n-1)$		$\pi_3 = \mathbb{Z}/28$

6.14. $h_0 = 1$ AND K_1 OF TYPE $\mathfrak{g}_2$

Then $H_1 \subseteq \mathrm{G}_2$ is a subgroup of type $\mathfrak{a}_1$ with large centralizer. The only possibilities are the two copies of $\mathrm{SU}(2)$ in $\mathrm{SO}(4) \subseteq \mathrm{G}_2$. In particular, $H_1 \cong \mathrm{SU}(2)$. For K_2 we have the possibilities $\mathrm{SU}(3)$, $\mathrm{Sp}(2)$, and G_2. We may write the map $H = H_1 \cdot H_0 \longrightarrow G = K_1 \times \mathrm{G}_2$ as $ab \longmapsto (\psi(a)\phi_1(b), \phi(b))$, where $\psi(a)\phi_1(b) \in \mathrm{SO}(4) \subseteq \mathrm{G}_2$ and $\phi : \mathrm{SU}(2) \longrightarrow G$. Thus $\{\psi, \phi_1\} = \{{}^{\mathbb{R}}\rho_{2\lambda_1}, {}^{\mathbb{R}}\rho_{2\lambda_1} + {}^{\mathbb{R}}\rho_{\lambda_1}\}$. It follows that the map $\pi_3(H) \longrightarrow \pi_3(G)$ is represented by the matrix

$$\begin{pmatrix} j_\psi & j_{\phi_1} \\ 0 & j_\phi \end{pmatrix}.$$

Note that $j_{({}^{\mathbb{R}}\rho_{\lambda_1})} = 1$ and $j_{({}^{\mathbb{R}}\rho_{2\lambda_1} + {}^{\mathbb{R}}\rho_{\lambda_1})} = 3$. We denote equivalence of matrices over $\mathbb{Z}$ by $\sim_{\mathbb{Z}}$.

$\boxed{K_2 = \mathrm{SU}(3)}$ Then we have the two possibilities $\phi = \rho_{\lambda_1}, \rho_{2\lambda_1}$. We obtain the matrices

$$\begin{pmatrix} 1 & 3 \\ 0 & 1 \end{pmatrix} \sim_{\mathbb{Z}} \begin{pmatrix} 1 & 0 \\ 0 & 1 \end{pmatrix} \text{ and } \begin{pmatrix} 3 & 1 \\ 0 & 1 \end{pmatrix} \sim_{\mathbb{Z}} \begin{pmatrix} 1 & 0 \\ 0 & 3 \end{pmatrix} \qquad (\phi = \rho_{\lambda_1})$$

$$\begin{pmatrix} 1 & 3 \\ 0 & 4 \end{pmatrix} \sim_{\mathbb{Z}} \begin{pmatrix} 1 & 0 \\ 0 & 4 \end{pmatrix} \text{ and } \begin{pmatrix} 3 & 1 \\ 0 & 4 \end{pmatrix} \sim_{\mathbb{Z}} \begin{pmatrix} 1 & 0 \\ 0 & 12 \end{pmatrix} \qquad (\phi = \rho_{2\lambda_1})$$

$\boxed{K_2 = \mathrm{Sp}(2)}$ The three possibilities for ϕ are ${}^{\mathbb{H}}\rho_{\lambda_1}, 2 \cdot {}^{\mathbb{H}}\rho_{\lambda_1}, {}^{\mathbb{H}}\rho_{3\lambda_1}$. Thus we have the matrices

$$\begin{pmatrix} 1 & 3 \\ 0 & 1 \end{pmatrix} \sim_{\mathbb{Z}} \begin{pmatrix} 1 & 0 \\ 0 & 1 \end{pmatrix} \text{ and } \begin{pmatrix} 3 & 1 \\ 0 & 1 \end{pmatrix} \sim_{\mathbb{Z}} \begin{pmatrix} 1 & 0 \\ 0 & 3 \end{pmatrix} \qquad (\phi = {}^{\mathbb{H}}\rho_{\lambda_1})$$

$$\begin{pmatrix} 1 & 3 \\ 0 & 2 \end{pmatrix} \sim_{\mathbb{Z}} \begin{pmatrix} 1 & 0 \\ 0 & 2 \end{pmatrix} \text{ and } \begin{pmatrix} 3 & 1 \\ 0 & 2 \end{pmatrix} \sim_{\mathbb{Z}} \begin{pmatrix} 1 & 0 \\ 0 & 6 \end{pmatrix} \qquad (\phi = 2 \cdot {}^{\mathbb{H}}\rho_{\lambda_1})$$

$$\begin{pmatrix} 1 & 3 \\ 0 & 10 \end{pmatrix} \sim_{\mathbb{Z}} \begin{pmatrix} 1 & 0 \\ 0 & 10 \end{pmatrix} \text{ and } \begin{pmatrix} 3 & 1 \\ 0 & 10 \end{pmatrix} \sim_{\mathbb{Z}} \begin{pmatrix} 1 & 0 \\ 0 & 30 \end{pmatrix} \qquad (\phi = {}^{\mathbb{H}}\rho_{3\lambda_1})$$

$\boxed{K_2 = \mathrm{G}_2}$ In this case we have the possibilities $\phi = {}^{\mathbb{R}}\rho_{\lambda_1}, {}^{\mathbb{R}}\rho_{\lambda_1} + {}^{\mathbb{R}}\rho_{2\lambda_1}, 2 \cdot {}^{\mathbb{R}}\rho_{2\lambda_1}, {}^{\mathbb{R}}\rho_{6\lambda_1}$. Accordingly, the matrices are as follows.

$$\begin{pmatrix} 1 & 3 \\ 0 & 1 \end{pmatrix} \sim_{\mathbb{Z}} \begin{pmatrix} 1 & 0 \\ 0 & 1 \end{pmatrix} \text{ and } \begin{pmatrix} 3 & 1 \\ 0 & 1 \end{pmatrix} \sim_{\mathbb{Z}} \begin{pmatrix} 1 & 0 \\ 0 & 3 \end{pmatrix} \qquad (\phi = {}^{\mathbb{R}}\rho_{\lambda_1})$$

$$\begin{pmatrix} 1 & 3 \\ 0 & 3 \end{pmatrix} \sim_{\mathbb{Z}} \begin{pmatrix} 1 & 0 \\ 0 & 3 \end{pmatrix} \text{ and } \begin{pmatrix} 3 & 1 \\ 0 & 3 \end{pmatrix} \sim_{\mathbb{Z}} \begin{pmatrix} 1 & 0 \\ 0 & 9 \end{pmatrix} \qquad (\phi = {}^{\mathbb{R}}\rho_{\lambda_1} + {}^{\mathbb{R}}\rho_{2\lambda_1})$$

$$\begin{pmatrix} 1 & 3 \\ 0 & 4 \end{pmatrix} \sim_{\mathbb{Z}} \begin{pmatrix} 1 & 0 \\ 0 & 4 \end{pmatrix} \text{ and } \begin{pmatrix} 3 & 1 \\ 0 & 4 \end{pmatrix} \sim_{\mathbb{Z}} \begin{pmatrix} 1 & 0 \\ 0 & 12 \end{pmatrix} \qquad (\phi = 2 \cdot {}^{\mathbb{R}}\rho_{\lambda_1})$$

$$\begin{pmatrix} 1 & 3 \\ 0 & 28 \end{pmatrix} \sim_{\mathbb{Z}} \begin{pmatrix} 1 & 0 \\ 0 & 28 \end{pmatrix} \text{ and } \begin{pmatrix} 3 & 1 \\ 0 & 28 \end{pmatrix} \sim_{\mathbb{Z}} \begin{pmatrix} 1 & 0 \\ 0 & 84 \end{pmatrix} \qquad (\phi = {}^{\mathbb{R}}\rho_{6\lambda_1})$$

K_1 of type G_2

K_2	(ψ, ϕ_1, ϕ)	(n_1, n_2)	Remarks
$\mathrm{SU}(3)$	$({}^{\mathbb{R}}\rho_{\lambda_1}, {}^{\mathbb{R}}\rho_{\lambda_1} + {}^{\mathbb{R}}\rho_{2\lambda_1}, \rho_{\lambda_1})$	$(5, 11)$	$\pi_5 = \mathbb{Z}/2$
$\mathrm{SU}(3)$	$({}^{\mathbb{R}}\rho_{\lambda_1} + {}^{\mathbb{R}}\rho_{2\lambda_1}, {}^{\mathbb{R}}\rho_{\lambda_1}, \rho_{\lambda_1})$	$(5, 11)$	$\pi_3 = \mathbb{Z}/4$
$\mathrm{SU}(3)$	$({}^{\mathbb{R}}\rho_{\lambda_1}, {}^{\mathbb{R}}\rho_{\lambda_1} + {}^{\mathbb{R}}\rho_{2\lambda_1}, 2\rho_{\lambda_1})$	$(5, 11)$	$\pi_3 = \mathbb{Z}/3$
$\mathrm{SU}(3)$	$({}^{\mathbb{R}}\rho_{\lambda_1} + {}^{\mathbb{R}}\rho_{2\lambda_1}, {}^{\mathbb{R}}\rho_{\lambda_1}, 2\rho_{\lambda_1})$	$(5, 11)$	$\pi_3 = \mathbb{Z}/12$
$\mathrm{Sp}(2)$	$({}^{\mathbb{R}}\rho_{\lambda_1}, {}^{\mathbb{R}}\rho_{\lambda_1} + {}^{\mathbb{R}}\rho_{2\lambda_1}, {}^{\mathbb{H}}\rho_{\lambda_1})$	$(7, 11)$	$\pi_5 = \mathbb{Z}/2$
$\mathrm{Sp}(2)$	$({}^{\mathbb{R}}\rho_{\lambda_1} + {}^{\mathbb{R}}\rho_{2\lambda_1}, {}^{\mathbb{R}}\rho_{\lambda_1}, {}^{\mathbb{H}}\rho_{\lambda_1})$	$(7, 11)$	$\pi_3 = \mathbb{Z}/3$
$\mathrm{Sp}(2)$	$({}^{\mathbb{R}}\rho_{\lambda_1}, {}^{\mathbb{R}}\rho_{\lambda_1} + {}^{\mathbb{R}}\rho_{2\lambda_1}, 2 \cdot {}^{\mathbb{H}}\rho_{\lambda_1})$	$(7, 11)$	$\pi_5 = \mathbb{Z}/2$
$\mathrm{Sp}(3)$	$({}^{\mathbb{R}}\rho_{\lambda_1} + {}^{\mathbb{R}}\rho_{2\lambda_1}, {}^{\mathbb{R}}\rho_{\lambda_1}, 2 \cdot {}^{\mathbb{H}}\rho_{\lambda_1})$	$(7, 11)$	$\pi_3 = \mathbb{Z}/6$
$\mathrm{Sp}(2)$	$({}^{\mathbb{R}}\rho_{\lambda_1}, {}^{\mathbb{R}}\rho_{\lambda_1} + {}^{\mathbb{R}}\rho_{2\lambda_1}, {}^{\mathbb{H}}\rho_{3\lambda_1})$	$(7, 11)$	$\pi_5 = \mathbb{Z}/10$
$\mathrm{Sp}(3)$	$({}^{\mathbb{R}}\rho_{\lambda_1} + {}^{\mathbb{R}}\rho_{2\lambda_1}, {}^{\mathbb{R}}\rho_{\lambda_1}, {}^{\mathbb{H}}\rho_{3\lambda_1})$	$(7, 11)$	$\pi_3 = \mathbb{Z}/30$
G_2	$({}^{\mathbb{R}}\rho_{\lambda_1}, {}^{\mathbb{R}}\rho_{\lambda_1} + {}^{\mathbb{R}}\rho_{2\lambda_1}, {}^{\mathbb{R}}\rho_{\lambda_1})$	$(11, 11)$	$\pi_5 = \mathbb{Z}/2 \oplus \mathbb{Z}/2$
G_2	$({}^{\mathbb{R}}\rho_{\lambda_1} + {}^{\mathbb{R}}\rho_{2\lambda_1}, {}^{\mathbb{R}}\rho_{\lambda_1}, {}^{\mathbb{R}}\rho_{\lambda_1})$	$(11, 11)$	$\pi_3 = \mathbb{Z}/3$
G_2	$({}^{\mathbb{R}}\rho_{\lambda_1}, {}^{\mathbb{R}}\rho_{\lambda_1} + {}^{\mathbb{R}}\rho_{2\lambda_1}, {}^{\mathbb{R}}\rho_{\lambda_1} + {}^{\mathbb{R}}\rho_{2\lambda_1})$	$(11, 11)$	$\pi_3 = \mathbb{Z}/3$
G_2	$({}^{\mathbb{R}}\rho_{\lambda_1} + {}^{\mathbb{R}}\rho_{2\lambda_1}, {}^{\mathbb{R}}\rho_{\lambda_1}, {}^{\mathbb{R}}\rho_{\lambda_1} + {}^{\mathbb{R}}\rho_{2\lambda_1})$	$(11, 11)$	$\pi_3 = \mathbb{Z}/9$
G_2	$({}^{\mathbb{R}}\rho_{\lambda_1}, {}^{\mathbb{R}}\rho_{\lambda_1} + {}^{\mathbb{R}}\rho_{2\lambda_1}, 2 \cdot {}^{\mathbb{R}}\rho_{2\lambda_1})$	$(11, 11)$	$\pi_3 = \mathbb{Z}/4$
G_2	$({}^{\mathbb{R}}\rho_{\lambda_1} + {}^{\mathbb{R}}\rho_{2\lambda_1}, {}^{\mathbb{R}}\rho_{\lambda_1}, 2 \cdot {}^{\mathbb{R}}\rho_{2\lambda_1})$	$(11, 11)$	$\pi_3 = \mathbb{Z}/12$
G_2	$({}^{\mathbb{R}}\rho_{\lambda_1}, {}^{\mathbb{R}}\rho_{\lambda_1} + {}^{\mathbb{R}}\rho_{2\lambda_1}, {}^{\mathbb{R}}\rho_{6\lambda_1})$	$(11, 11)$	$\pi_3 = \mathbb{Z}/28$
G_2	$({}^{\mathbb{R}}\rho_{\lambda_1} + {}^{\mathbb{R}}\rho_{2\lambda_1}, {}^{\mathbb{R}}\rho_{\lambda_1}, {}^{\mathbb{R}}\rho_{6\lambda_1})$	$(11, 11)$	$\pi_3 = \mathbb{Z}/84$

THEOREM 6.15. *Let G/H be a 1-connected homogeneous space of a compact connected Lie group G. Assume that*

$$\mathbf{H}^\bullet(G/H) = \bigwedge\nolimits_{\mathbb{Z}}(u, v)$$

is an exterior algebra on two generators of odd degrees $\deg(u) = n_1, \deg(v) = n_2$, *with $n_1, n_2 \geq 3$. Assume moreover that the action is irreducible, and not split, and that G is not almost simple. Then (G, H) is one of the following pairs.*

G semisimple

$G = K_1 \times K_2$	$H = H_1 \cdot H_0$	$\mathrm{Cen}_G(H)^\circ$	G/H
$\mathrm{Sp}(n) \times \mathrm{SU}(3)$, $n \geq 2$	$\mathrm{Sp}(n-1) \cdot \mathrm{Sp}(1)$	$\mathrm{U}(1)$	$\mathbb{S}^{4n-1} \times \mathbb{S}^5$
$\mathrm{Sp}(n) \times \mathrm{Sp}(2)$, $n \geq 2$	$\mathrm{Sp}(n-1) \cdot \mathrm{Sp}(1)$	$\mathrm{Sp}(1)$	$\mathbb{S}^{4n-1} \times \mathbb{S}^7$

In both series, the group H_0 is the image of $\mathrm{Sp}(1)$ *in* $\mathrm{Cen}_G(H_1)^\circ = \mathrm{Sp}(1) \times K_2$ *under the diagonal embedding $h \longmapsto (h, h)$.* □

6.C. The non-split case (II): $\mathbf{H}^\bullet(X) = \mathbb{Z}[a]/(a^2) \otimes \bigwedge_{\mathbb{Z}}(w)$.

We use the same conventions as in the last section. This time, we have

$$\begin{aligned} 1 &= \mathrm{rk}(G) - \mathrm{rk}(H) \\ &= k_1 - h_1 + k_2 - h_2 - h_0 \\ 0 &\leq k_1 - h_1 - h_0 \\ 0 &\leq k_2 - h_2 - h_0, \end{aligned}$$

whence $h_0 = 1$. Thus, H_0 is either a 1-torus, or of type $\mathfrak{a}_1$, and $k_i - h_i - h_0 = 0$ for $i = 1, 2$.

LEMMA 6.16. *If $n_1 = 4$, H_0 is of type $\mathfrak{a}_1$, and (K_1, H_1) is one of the pairs in 6.9. Moreover, $(K_2, H_2) = (\mathrm{Sp}(2), \mathrm{Sp}(1))$.*

PROOF. From the exact sequence

$$\underbrace{\pi_2(G/H)}_{0} \longrightarrow \pi_1(H) \longrightarrow \pi_1(G) \longrightarrow \underbrace{\pi_1(G/H)}_{0}$$

we see that H is semisimple. Moreover,

$$\underbrace{\pi_4(G)}_{\text{finite}} \longrightarrow \underbrace{\pi_4(G/H)}_{\mathbb{Z}} \longrightarrow \pi_3(H) \longrightarrow \pi_3(G) \longrightarrow \underbrace{\pi_3(G/H)}_{0}$$

implies that $\mathrm{rk}(\pi_3(H)) = \mathrm{rk}(\pi_3(G)) + 1 = 3$. Thus, H is semisimple with three almost simple factors. One of them has to be H_0, hence H_0 is of type $\mathfrak{a}_1$.

Thus, $P^k_{H_i} \longleftarrow P^k_{K_i}$ is a surjection in all degrees $k \geq 5$ (and in fact also in degree 3, since the map induced on π_3 is not trivial); thus, (K_1, H_1) and (K_2, H_2) are among the pairs determined in 6.9. Moreover, $\mathrm{rk}(\pi_7(G/H)) \in \{1, 2\}$, cp. 3.11, hence $\mathrm{rk}(\pi_7(G)) - \mathrm{rk}(\pi_7(H)) = \mathrm{rk}(\pi_7(G/H)) \geq 1$. We may assume that $\mathrm{rk}(\pi_7(K_2)) - \mathrm{rk}(\pi_7(H_2)) \geq 1$. Thus $(K_2, H_2) = (\mathrm{Sp}(2), \mathrm{Sp}(1))$. □

LEMMA 6.17. *The case $n_1 \geq 6$ is not possible.*

PROOF. In this case

$$\mathrm{rk}(\pi_i(H)) = \mathrm{rk}(\pi_i(G)) = 2 \quad \text{for } i = 1, 2, 3,$$

hence H is semisimple with two almost simple factors. We may assume that $H_2 = 1$. Note that $h_0 = 1$, hence $k_2 = 1$. Therefore K_2 is of type $\mathfrak{a}_1$. But then $\mathrm{pr}_2(H_0) = K_2$, a contradiction to the irreducibility of the action. □

We classify the remaining possibilities.

6.18. $K_1 = \mathrm{Sp}(n)$, $n \geq 2$

Recall that there is a fibre bundle

$$\begin{array}{ccc} \mathbb{S}^{4n-1} = \mathrm{Sp}(n)/\mathrm{Sp}(n-1) & \longrightarrow & G/H \\ & & \downarrow \\ & & \mathrm{Sp}(2)/\mathrm{Sp}(1) \times \mathrm{Sp}(1) = \mathbb{S}^4. \end{array}$$

In fact, this bundle is the Whitney sum of n copies of the quaternionic Hopf bundle over $\mathbb{S}^4 = \mathbb{H}\mathrm{P}^1$.

6.19. $K_1 = \mathrm{G}_2$

There is a similar fibre bundle

$$\begin{array}{ccc} \mathrm{G}_2/H_1 & \longrightarrow & G/H \\ & & \downarrow \\ & & \mathrm{Sp}(2)/\mathrm{Sp}(1) \times \mathrm{Sp}(1) = \mathbb{S}^4. \end{array}$$

By 6.5, the fibre has the same $\mathbb{Z}/2$-cohomology as the Stiefel manifold $V_2(\mathbb{R}^7)$. Therefore the $\mathbb{Z}/2$-Leray Serre spectral sequence of the bundle collapses, and thus

$$\mathbf{H}^\bullet(G/H;\mathbb{Z}/2) \cong \bigwedge\nolimits_{\mathbb{Z}/2}(x_5, x_7) \otimes (\mathbb{Z}/2)[x_4, x_6]/(x_4^2, x_6^2).$$

THEOREM 6.20. *Let G/H be a compact 1-connected homogeneous space of a compact connected Lie group G. Assume that G/H has the same integral cohomology as $\mathbb{S}^{n_1} \times \mathbb{S}^{n_2}$, for $n_1 \geq 4$ even and $n_2 > n_1$ odd. If the action of G is irreducible and not split, and if G is not simple, then $n_1 = 4$ and (G, H) is one of the following pairs.*

G semisimple

$G = G_1 \times G_2$	$H = (H_1 \times H_2) \cdot H_0$	$\mathrm{Cen}_G(H)^\circ$	(n_1, n_2)	G/H
$\mathrm{Sp}(n) \times \mathrm{Sp}(2)$ $n \geq 2$	$(\mathrm{Sp}(n-1) \times \mathrm{Sp}(1)) \cdot \mathrm{Sp}(1)$	*1*	$(4, 4n-1)$	$S(n\eta_{\mathbb{H}})$

The group H_0 is the image of $\mathrm{Sp}(1)$ *in* $\mathrm{Cen}_G(H_1 \times H_2)^\circ = \mathrm{Sp}(1) \times \mathrm{Sp}(1)$ *under the diagonal embedding $h \longmapsto (h, h)$. The space G/H is the sphere bundle $S(n\eta_{\mathbb{H}})$ of the Whitney sum $n\eta_{\mathbb{H}}$ of n copies of the quaternionic Hopf bundle $\eta_{\mathbb{H}}$ over $\mathbb{S}^4$.* □

CHAPTER 7

Homogeneous compact quadrangles

The principal aim of topological geometry is to classify reasonable geometries in terms of their automorphism groups. 'Reasonable' geometries are linear spaces (e.g. projective, affine or hyperbolic geometries, or, more generally, stable planes), circle geometries (Laguerre or Möbius geometries) and finally Tits buildings, to us the most important class of geometries.

The fundamental theorem of projective geometry asserts that a projective space of rank at least 3 (i.e. a projective space which is not a projective plane) is coordinatized by a field or skew field. The key step in the proof is to show that such a projective space satisfies the Desargues condition. There is a similar result due to Tits [**106**] for spherical buildings. An irreducible spherical building of rank at least 3 satisfies the so-called Moufang condition; furthermore, all irreducible spherical Moufang buildings of rank at least 2 were determined by Tits [**106**] and Tits-Weiss [**108**].

Moufang buildings are rather important in many branches of mathematics. Just to mention a few examples, differential geometers use them in the proof of the rigidity theorems of Mostow [**75**], Gromov [**3**], Kleiner-Leeb [**50**] and Leeb [**65**], and in the classification of isoparametric submanifolds, see Thorbergsson [**101**]; they play a rôle in connection with S-arithmetic groups and group cohomology, see Rohlfs-Springer [**83**] and Abramenko [**1**], and in model theory in connection with simple superstable groups, see Borovik-Nesin [**14**] and Kramer-Tent-Van Maldeghem [**60**].

In Tits' classification, the assumption that the building has rank at least 3 cannot be dropped; there exist uncountably many 'wild' buildings of rank 2, even with large automorphism groups, see Tits [**107**] and Tent [**99**]. Thus it is desirable to have a criterion which ensures that an infinite building of rank 2 is Moufang. Indeed, such a criterion exists in the topological category.

Theorem A *A compact connected irreducible spherical building of rank at least* 2 *with a flag transitive automorphism group is Moufang.*

This is proved in a series of papers by Grundhöfer-Knarr-Kramer [**38**] [**39**] [**40**]. Note that it suffices to consider spherical buildings of rank 2 to prove the theorem. No similar classification is presently known for finite or zero-dimensional (totally disconnected) spherical buildings of rank 2. (The finite Moufang buildings are closely related to finite simple groups; totally disconnected buildings appear rather naturally in connection with valuations and algebraic groups over local fields.) In the course of the proof, the closed connected flag transitive groups are also explicitly determined. It turns out that such a group G is either the little projective group of the building (the group generated by the root groups) or a certain compact subgroup of the little projective group. The compact connected flag transitive groups were determined by Eschenburg-Heintze [**31**]; all closed connected flag transitive groups are determined in Grundhöfer-Knarr-Kramer [**38**] [**39**] [**40**].

In this chapter we extend Theorem A to compact connected buildings with a less transitive automorphism group, that is, an automorphism group which is only transitive on vertices of a certain type. By Tits' results about buildings of higher rank, we have only to consider buildings of rank 2. Spherical buildings of rank two are commonly called generalized polygons. In building terminology, they are certain 1-dimensional numbered simplicial complexes, i.e. bipartite graphs: there are precisely two types of vertices in such a building which we call points and lines. The axioms of a building are symmetric in the sense that we might as well call the lines points and the points lines. Thus, there is no loss in generality if one studies either point or line transitive actions. The main result of this chapter can be stated as follows.

Theorem B *A compact connected building of type C_2 with a point transitive automorphism group is Moufang, provided that the point space is 9-connected (7.33, 7.34).*

It is known that there exist line homogeneous compact connected generalized quadrangles with a 6-connected line space which are not Moufang, see Kramer [**55**]; in particular, point or line transitivity does not imply the Moufang property in general. This indicates already that the problem is more difficult than the flag transitive case considered in Grundhöfer-Knarr-Kramer [**38**] [**39**].

A result of Knarr [**51**] and the author [**54**] says that a finite dimensional compact connected polygon is a building of type A_2, C_2 or G_2, i.e. a projective plane, a generalized quadrangle (which is the same as a polar space of rank 2) or a generalized hexagon. A point transitive automorphism group implies that the topological dimension is finite, so this result can be applied. The three types of buildings then have to be considered separately. The A_2-case was done by Löwen and Salzmann about 20 years ago [**84**] [**66**].

Theorem C *A compact connected projective plane or generalized hexagon with a point transitive automorphism group is Moufang, see Salzmann* [**84**], *Löwen* [**66**] *or Salzmann* et al. [**85**] *63.8, and Kramer* [**54**] *Ch. 5. A compact connected generalized quadrangle with a point transitive automorphism group is Moufang, provided that the point and line space have the same dimension, see Kramer* [**54**] *Ch. 5.*

Combining Theorems A, B and C, we obtain the following result.

Theorem D *Let Δ be an irreducible spherical compact connected building of rank at least* 2. *Assume that the automorphism group of Δ acts transitively on one type of vertices. If the building is of type C_2, assume in addition that either (a) the vertex set in question is* 9*-connected, or (b) the two vertex sets of the building have the same dimension, or (c) that the action is chamber transitive. Then Δ is the Moufang building associated to a non-compact real simple Lie group.*

Before I comment on the proof of Theorem B, I would like to mention some recent developments. Biller studied in his Ph.D. thesis [**4**] (not necessarily transitive) compact group actions on compact generalized quadrangles. In particular, he characterized the hermitian quadrangles over the quaternions in terms of the size of their automorphism groups. His work contains many new results and will certainly be an important source for further research in this direction. We use some of his results in Section 4 of this chapter; my original proof of Theorem 7.31 (as given in [**56**] Section 7.D) is incorrect, and the new proof given here relies on Biller's ideas. Recently, Bletz and Wolfrom started to investigate point transitive actions on compact connected (m_1, m_2)-quadrangles, with $m_1 = 1, 2$. Finally, Immervoll [**48**] proved that all isoparametric hypersurfaces with four distinct principal curvatures are generalized quadrangles (cp. Chapter 8). This has been a difficult open problem for quite some time. His proof is a clever combination of transversality arguments and the algebraic machinery developed by Dorfmeister-Neher in the early 80s.

Some remarks on the proof of Theorem B. In order to prove Theorem B, one has to consider generalized quadrangles whose point and line spaces have different dimensions. The cohomology ring of the point space of such a quadrangle is known (Kramer [**54**], Strauß [**92**]), and it turns out that it looks in most cases like the cohomology of a product of spheres. Hence we can apply our classification of homogeneous spaces.

Not every homogeneous space in our list 3.15 corresponds to a compact quadrangle. The point stabilizer cannot have large normal subgroups, cp. 7.21 below. This excludes for example the compact symmetric space $\mathrm{E}_6/\mathrm{F}_4$.

Besides the Stiefel manifold $V_2(\mathbb{F}^n)$, $\mathbb{F} = \mathbb{R}, \mathbb{C}, \mathbb{H}$ we have to consider certain homogeneous sphere bundles over spheres, and also products of homogeneous spheres.

For the Stiefel manifolds we show the following: if the dimension of the corresponding vector space $\mathbb{F}^n$ is at least 5 (at least 4 for $\mathbb{F} = \mathbb{H}$), then there is a unique quadrangle compatible with the transitive group action which 'lives' on the Stiefel manifold.

A similar result is proved for one of the three series of homogeneous sphere bundles. As mentioned before, this proof relies on Biller's thesis [**4**].

For the products of homogeneous spheres we show that one of the two factors has to be $\mathbb{S}^3$ with the regular Sp(1)-action, and we obtain some restrictions on the other factor. The examples of Ferus-Karcher-Münzner [**34**] show that there is at least one family of non-Moufang quadrangles which 'lives' on such homogeneous products of spheres, see Example 8.13 in Chapter 8.

The relevant results about topological generalized quadrangles can be found in Grundhöfer-Knarr [**37**], Grundhöfer-Van Maldeghem [**42**], Grundhöfer-Löwen [**41**], Grundhöfer-Knarr-Kramer [**38**] [**39**], Knarr [**51**], Kramer-Van Maldeghem [**59**] and Kramer [**57**] [**58**], Schroth [**88**], and in particular in Kramer [**54**]; Van Maldeghem's monograph [**110**] is the authoritative source for geometric properties of generalized polygons; Chapter 9 in his book summarizes many results about topological polygons.

7.A. Generalized quadrangles and group actions

A *point-line geometry* is a triple

$$\mathfrak{G} = (\mathcal{P}, \mathcal{L}, \mathcal{F})$$

consisting of a set $\mathcal{P}$ of *points*, a set $\mathcal{L}$ of *lines* and a set $\mathcal{F} \subseteq \mathcal{P} \times \mathcal{L}$ of *flags*. If a pair $(p, \ell) \in \mathcal{F}$ is a flag, then p and ℓ are called *incident*; one also says that the line ℓ passes through the point p, or that p lies on ℓ. (One can turn $\mathfrak{G}$ into a *bipartite graph* (V, E) with vertex set $V = \mathcal{P} \cup \mathcal{L}$ and edge set $E = \{\{p, \ell\} | \ (p, \ell) \in \mathcal{F})\}$; this is the building point of view.) A *k-chain* joining $x_0, x_k \in \mathcal{P} \cup \mathcal{L}$ is a sequence $(x_0, \dots, x_k) \in (\mathcal{P} \cup \mathcal{L})^{k+1}$ with the property that x_i is incident with x_{i-1} for $1 \leq i \leq k$. The *distance* of x_0 and x_k is

$$d(x_0, x_k) = k$$

if there is a k-chain joining x_0 and x_k, but no j-chain joining x_0 and x_k for $j < k$. Put

$$D_k(x) = \{y \in \mathcal{P} \cup \mathcal{L} |\ d(x, y) = k\}.$$

A point-line geometry is called *thick* if $|D_1(x)| \geq 3$ for all $x \in \mathcal{P} \cup \mathcal{L}$. We put

$$x^\perp = \{x\} \cup D_2(x);$$

if x is a point, then this is the set of all points which have a line in common with x.

Automorphisms are defined in the obvious way; an automorphism is a permutation of the point and the line set which preserves the incidence relation. In graph theoretic terms, an automorphism is a graph automorphism which preserves the coloring of the bipartite graph.

7.1. Generalized quadrangles

A thick point-line geometry is called a *generalized quadrangle* if $\mathcal{F} \neq \mathcal{P} \times \mathcal{L}$, and if

$$|p^\perp \cap D_1(\ell)| = 1$$

holds for all pairs $(p, \ell) \in (\mathcal{P} \times \mathcal{L}) \setminus \mathcal{F}$. Then there is a unique 3-chain (p, h, q, ℓ) joining p and ℓ. Put

$$q = \mathrm{proj}_\ell p \quad \text{and} \quad h = \mathrm{proj}_p \ell.$$

The picture below shows the corresponding 3-chain.

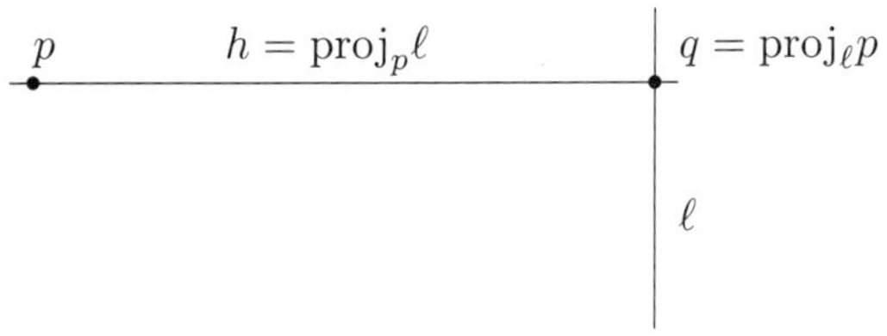

Note that a generalized quadrangle contains no digons (two point are joined by at most one line and two lines intersect in at most one point) and no triangles. In particular, a line ℓ is uniquely determined by the point row $D_1(\ell)$. For an example of a generalized quadrangle, see 7.8 below. *A generalized quadrangle is the same as a spherical building of type* C_2.

7.2. Group actions and reconstruction

An *action* of a group G on a generalized quadrangle $\mathfrak{G}$ is a homomorphism

$$G \longrightarrow \mathrm{Aut}(\mathfrak{G}).$$

Suppose that

Rec$_1$ G acts transitively on the point set $\mathcal{P}$.

The question is whether $\mathfrak{G}$ can be *reconstructed* from the action of G. Let $p \in \mathcal{P}$ and assume that

Rec$_2$ the stabilizer G_ℓ acts transitively on $D_1(\ell)$ for every $\ell \in D_1(p)$.

Then $D_1(\ell) = G_\ell \cdot p \subseteq \mathcal{P}$. A quadrangle satisfying **Rec**$_1$ and **Rec**$_2$ is an example of what Stroppel **[93]** **[94]** calls a *sketched geometry*. Suppose we know the collection

$$\mathcal{G} = \{G_\ell |\ \ell \in D_1(p)\}$$

of all stabilizers of lines passing through p. Then we know all point rows through p; they are the sets $G_\ell / G_\ell \cap G_p \subseteq G/G_p$. The collection of their G-translates can be canonically identified with the line set $\mathcal{L}$. Thus, the quadrangle is uniquely determined by the triple

$$(G, G_p, \mathcal{G}).$$

In fact, put

$$\mathcal{P}' = G/G_p \quad \text{and} \quad \mathcal{L}' = \bigcup\{gHG_p | \ g \in G, \ H \in \mathcal{G}\}$$

and define the incidence as inclusion '$\subseteq$'. It is easy to show that the resulting geometry $(\mathcal{P}', \mathcal{L}', \subseteq)$ is G-isomorphic to the original quadrangle $\mathfrak{G}$. See Stroppel [**93**] [**94**] for more results in this direction.

We will also use the following elementary result.

LEMMA 7.3. *Let $\mathfrak{G}$ be a point-line geometry, with the property that every line ℓ is uniquely determined by its point row $D_1(\ell)$. Assume that $d(x, y) < \infty$ holds for all $x, y \in \mathcal{P} \cup \mathcal{L}$, and that $G \subseteq \mathrm{Aut}(\mathfrak{G})$ acts transitively on $\mathcal{P}$. Let $p \in \mathcal{P}$ and let $N \subseteq G_p$ be a subgroup with the following two properties:*

(i) *N acts trivially on $D_1(p)$ and on $D_1(\ell)$, for all $\ell \in D_1(p)$ (so N fixes $p^\perp$ pointwise).*
(ii) *If $gNg^{-1} \subseteq G_p$, for some $g \in G$, then $gNg^{-1} = N$.*

Then $N = 1$.

PROOF. Let q be a point which is collinear with p, i.e. $q \in D_2(p)$. Since G acts transitively on $\mathcal{P}$, the stabilizer G_q is conjugate to G_p, say $G_q = gG_pg^{-1}$. Then gNg^{-1} fixes $D_2(q)$ elementwise; in particular, $gNg^{-1} \subseteq G_p$. Thus $N = gNg^{-1}$. It follows inductively that N is normal in G, and thus $N = 1$, because the action of G on $\mathcal{P}$ is effective (here we use that assumption that lines are determined by their point rows). $\square$

7.B. Compact quadrangles

Suppose that $\mathfrak{G} = (\mathcal{P}, \mathcal{L}, \mathcal{F})$ is a generalized quadrangle, and that the sets $\mathcal{P}$ and $\mathcal{L}$ carry Hausdorff topologies. If the map $(p, \ell) \longmapsto (q, h) = (\mathrm{proj}_\ell p, \mathrm{proj}_p \ell)$ is continuous on $(\mathcal{P} \times \mathcal{L}) \setminus \mathcal{F}$, then $\mathfrak{G}$ is called a *topological quadrangle*; if $\mathcal{P}$ and $\mathcal{L}$ are in addition compact, then $\mathfrak{G}$ is called a *compact quadrangle*.

PROPOSITION 7.4. *Let $\mathfrak{G}$ be a generalized quadrangle, and suppose that $\mathcal{P}$ and $\mathcal{L}$ are compact Hausdorff spaces. Then $\mathfrak{G}$ is a compact quadrangle if and only if $\mathcal{F} \subseteq \mathcal{P} \times \mathcal{L}$ is closed (or, equivalently, compact).*

PROOF. See Grundhöfer-Van Maldeghem [**42**] 2.1. $\square$

Concerning the set theoretic topology of a compact quadrangle, the following result is important.

PROPOSITION 7.5. *Let $\mathfrak{G}$ be a compact quadrangle. The point space $\mathcal{P}$ and the line space $\mathcal{L}$ are second countable, separable and metrizable. If the point space $\mathcal{P}$ is connected, then $\mathcal{P}$, $\mathcal{L}$, every point row $D_1(\ell)$ and every line pencil $D_1(p)$ is connected and locally contractible.*

PROOF. See Grundhöfer-Knarr [**37**] 3.1 and 4.1, or Grundhöfer-Knarr-Kramer [**38**] 1.5 and 1.6. $\square$

Recall the definition of the *covering dimension* $\dim(X)$ of a normal space X. If every finite open covering of X has a refinement such that every point of X is contained in at most $n+1$ sets of the refinement, then $\dim(X) \leq n$, and $\dim(X) = n$ if X has at most dimension n, but not dimension $n-1$. Equivalently, $\dim(X) \leq n$ holds if and only if the extension problem

$$\begin{array}{ccc} A & \xrightarrow{\alpha} & \mathbb{S}^n \\ \big\downarrow & \nearrow & \\ X & & \end{array}$$

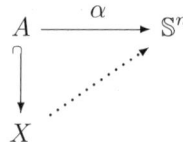

has a solution for every map α defined on a closed subspace $A \subseteq X$. There are other topological notions of dimension, most notably the small or large inductive dimension, and the cohomological dimension. See Salzmann *et al.* [**85**] 92 for a discussion of dimension functions and further references. For 'good' spaces, these notions of dimension agree; in particular, an n-manifold has dimension n. More generally, an *integral ENR n-manifold* has dimension n.

DEFINITION 7.6. Let X be an ENR (euclidean neighborhood retract, cp. Dold [**24**] IV.8 for properties of such spaces). If there exists a number n such that

$$\mathbf{H}_k(X, X \setminus \{x\}) \cong \begin{cases} \mathbb{Z} & \text{for } k = n \\ 0 & \text{for } k \neq n \end{cases}$$

holds for all $x \in X$, then X is called an *integral ENR manifold.* Integral ENR manifolds are *generalized manifolds*, i.e. cm_Rs and hm_Rs (for any integral domain R) as studied in Bredon [**17**], and it can be shown that $\dim(X) = n$. This is due to the fact that an ENR is locally contractible (so singular homology coincides with Borel-Moore homology) and a result of Bredon [**17**]. (An ENR is locally contractible and in particular clc_R^∞; since $\dim(X)$ is finite, $\dim_R(X)$ is also finite. The local homology groups in Borel-Moore homology are isomorphic to the local homology groups in singular homology. The result now follows from Theorem V.16.8 in Bredon [**17**].) Integral ENR manifolds are a very convenient class of spaces, because they satisfy all standard topological assumptions in the theory of compact transformation groups.

Note also that an ANR of finite covering dimension is the same as an ENR.

A compact quadrangle is called *finite dimensional* if the covering dimension of $\mathcal{P}$ is finite and positive, or, equivalently, if the covering dimension of $\mathcal{L}$ is finite and positive. The following theorem summarizes the most important topological properties of compact connected finite dimensional quadrangles.

THEOREM 7.7. *Suppose that $\mathfrak{G}$ is a finite dimensional compact quadrangle. Let $(p, \ell) \in \mathcal{F}$ and put $(m_1, m_2) = (\dim(D_1(\ell)), \dim(D_1(p)))$.*

Then the following hold.

(i) *The spaces $\mathcal{P}$, $\mathcal{L}$ and $\mathcal{F}$ are ENRs (euclidean neighborhood retracts) and in particular ANRs (absolute neighborhood retracts).*
(ii) *The spaces $\mathcal{P}$, $\mathcal{L}$ and $\mathcal{F}$, as well as the point rows and the line pencils are integral (locally and globally homogeneous) ENR-manifolds.*
(iii) *The inclusions $\{p\} \subseteq D_1(\ell) \subseteq p^\perp \subseteq \mathcal{P}$ are cofibrations for every pair (p, ℓ) (and dually).*

(iv) *There are homotopy equivalences*

$$\begin{aligned} D_1(\ell) &\simeq \mathbb{S}^{m_1} \\ p^\perp/D_1(\ell) &\simeq \mathbb{S}^{m_1+m_2} \\ \mathcal{P}/p^\perp &\simeq \mathbb{S}^{2m_1+m_2} \end{aligned}$$

(v) $D_1(\ell)$, $p^\perp$ *and* $\mathcal{P}$ *are* $(m_1 - 1)$*-connected.*
(vi) *Either* $m_1 = m_2 \in \{1, 2, 4\}$, *or* $1 \in \{m_1, m_2\}$, *or* $m_1 + m_2$ *is odd.*

PROOF. This is proved in Knarr [**51**], Grundhöfer-Knarr [**37**] Sec. 4 and in the present generality in Kramer [**54**] Thm. 3.3.6. □

The numbers (m_1, m_2) are called the *topological parameters* of $\mathfrak{G}$. A compact connected finite dimensional quadrangle with parameters (m_1, m_2) is called a *compact connected* (m_1, m_2)*-quadrangle* for short.

7.8. EXAMPLE Let $\mathbb{F} = \mathbb{R}, \mathbb{C}, \mathbb{H}$ and consider the hermitian form in $n + 1$ variables over $\mathbb{F}$

$$h(x, y) = -\bar{x}_0 y_0 - \bar{x}_1 y_1 + \sum_{k=2}^{n} \bar{x}_k y_k.$$

Let $\mathcal{P}$ denote the collection of all 1-dimensional totally isotropic subspaces, and $\mathcal{L}$ the collection of all 2-dimensional totally isotropic subspaces of $\mathbb{F}^{n+1}$ (a subspace V of $\mathbb{F}^{n+1}$ is called *totally isotropic* if the form h vanishes identically on V). If $n \geq 4$, or if $\mathbb{F} \neq \mathbb{R}$ and $n = 3$, then

$$\mathsf{H}_n(\mathbb{F}, \mathbb{R}) = (\mathcal{P}, \mathcal{L}, \subseteq)$$

is a compact connected quadrangle, the *standard hermitian quadrangle* over $\mathbb{F}^{n+1}$; one puts $\mathsf{Q}_n(\mathbb{R}) = \mathsf{H}_n(\mathbb{R}, \mathbb{R})$ (our notation is essentially the same as in Van Maldeghem [**110**]). Let $d = \dim_{\mathbb{R}}(\mathbb{F}) = 1, 2, 4$. The quadrangle $\mathsf{H}_n(\mathbb{F}, \mathbb{R})$ has parameters $(d, d(n-2) - 1)$. Note that there are natural inclusions

$$\begin{array}{ccccccccc}
 & & \mathsf{Q}_4(\mathbb{R}) & \hookrightarrow & \mathsf{Q}_5(\mathbb{R}) & \hookrightarrow & \mathsf{Q}_6(\mathbb{R}) & \hookrightarrow & \cdots \\
 & & \downarrow & & \downarrow & & \downarrow & & \\
\mathsf{H}_3(\mathbb{C}, \mathbb{R}) & \hookrightarrow & \mathsf{H}_4(\mathbb{C}, \mathbb{R}) & \hookrightarrow & \mathsf{H}_5(\mathbb{C}, \mathbb{R}) & \hookrightarrow & \mathsf{H}_6(\mathbb{C}, \mathbb{R}) & \hookrightarrow & \cdots \\
\downarrow & & \downarrow & & \downarrow & & \downarrow & & \\
\mathsf{H}_3(\mathbb{H}, \mathbb{R}) & \hookrightarrow & \mathsf{H}_4(\mathbb{H}, \mathbb{R}) & \hookrightarrow & \mathsf{H}_5(\mathbb{H}, \mathbb{R}) & \hookrightarrow & \mathsf{H}_6(\mathbb{H}, \mathbb{R}) & \hookrightarrow & \cdots
\end{array}$$

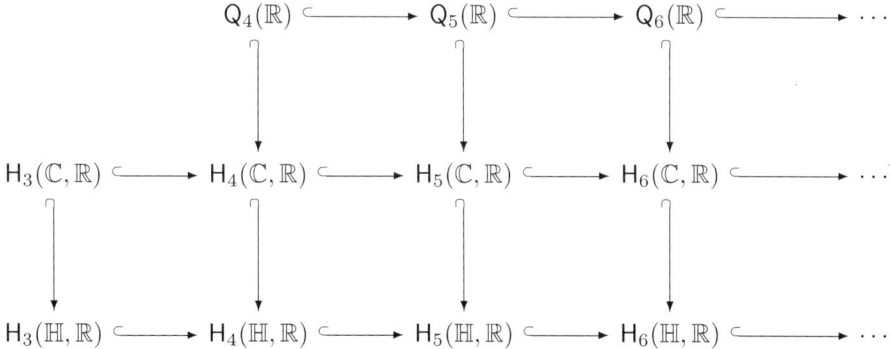

In the limit one obtains non-compact topological quadrangles with compact point rows and contractible infinite dimensional line pencils modeled on $\mathbb{S}^\infty$.

Let $\mathfrak{G} = (\mathcal{P}, \mathcal{L}, \mathcal{F})$ be a generalized quadrangle, let $\mathcal{P}' \subseteq \mathcal{P}$ and $\mathcal{L}' \subseteq \mathcal{L}$ be non-empty subsets. If $\mathfrak{G}' = (\mathcal{P}', \mathcal{L}', \mathcal{F} \cap (\mathcal{P}' \times \mathcal{L}'))$ is again a generalized quadrangle, then we call $\mathfrak{G}'$ a *subquadrangle* of $\mathfrak{G}$. If $D_1(\ell) \subseteq \mathcal{P}'$ holds for all $\ell \in \mathcal{L}'$, then the subquadrangle is called *full*. Here is an example of a full and a non-full subquadrangle.

$$\begin{array}{ccc} \mathsf{H}_3(\mathbb{C},\mathbb{R}) & \xhookrightarrow{\text{not full}} & \mathsf{H}_3(\mathbb{H},\mathbb{R}) \\ \big\downarrow \text{full} & & \\ \mathsf{H}_7(\mathbb{C},\mathbb{R}) & & \end{array}$$

The following result is proved in Kramer-Van Maldeghem [**59**] Thm. 4.1.

PROPOSITION 7.9. *Let $\mathfrak{G}' \subset \mathfrak{G}$ be a full, compact, and proper (i.e. $\mathfrak{G}' \neq \mathfrak{G}$) subquadrangle of the compact connected (m_1, m_2)-quadrangle $\mathfrak{G}$. Let (m_1, m_2') denote the parameters of $\mathfrak{P}'$. Then $m_2' \neq 0$ (so $\mathfrak{G}'$ is connected), and*

$$m_1 + m_2' \leq m_2.$$

PROOF. The idea of the proof is to show that $\mathfrak{G}'$ contains a compact ovoid $\mathcal{O}$ which injects into a line pencil of $\mathfrak{G}$. The dimension of such an ovoid is $m_1 + m_2'$, whence $m_1 + m_2' \leq m_2$. For details see Kramer-Van Maldeghem Thm. 4.1 [**59**]. The fact that the point rows of $\mathfrak{G}'$ are connected implies that the line pencils are also connected, whence $m_2' \neq 0$. □

The cohomology of finite dimensional quadrangles is known.

7.10. COHOMOLOGY OF FINITE DIMENSIONAL COMPACT QUADRANGLES

Let $\mathfrak{G}$ be a compact connected (m_1, m_2)-quadrangle. Then

$$\begin{aligned} \dim(\mathcal{F}) &= 2(m_1 + m_2) \\ \dim(\mathcal{P}) &= 2m_1 + m_2 \\ \dim(\mathcal{L}) &= 2m_2 + m_1. \end{aligned}$$

If $m_1 + m_2$ is odd, then

$$\begin{aligned} \mathbf{H}^\bullet(\mathcal{P}) &\cong \mathbf{H}^\bullet(\mathbb{S}^{m_1} \times \mathbb{S}^{m_1+m_2}) \\ \mathbf{H}^\bullet(\mathcal{L}) &\cong \mathbf{H}^\bullet(\mathbb{S}^{m_2} \times \mathbb{S}^{m_1+m_2}) \\ \mathbf{H}^\bullet(\mathcal{F}) &\cong \mathbf{H}^\bullet(\mathbb{S}^{m_1} \times \mathbb{S}^{m_2} \times \mathbb{S}^{m_1+m_2}). \end{aligned}$$

If $m_1 = 1$ and if $m_2 > 2$ is odd, then

$$\mathbf{H}^\bullet(\mathcal{L}; \mathbb{Q}) \cong \mathbf{H}^\bullet(\mathbb{S}^{2m_2+1}; \mathbb{Q}).$$

The group $\pi_{m_2}(\mathcal{L})$ is generated by the inclusion $\mathbb{S}^{m_2} \simeq D_1(p) \subseteq \mathcal{L}$. If $m_1 + m_2$ is odd, then $\pi_{m_1}(\mathcal{L}) \cong \mathbb{Z}$, and if m_2 is odd and $m_1 = 1$, then $\pi_{m_2}(\mathcal{L}) \cong \mathbb{Z}/2$.

These results are proved in Kramer [**54**] Ch. 3 and 6.4; they follow from Münzner [**77**], cp. also Strauß [**92**].

Let $\mathfrak{G}$ be a compact generalized quadrangle (not necessarily connected, not necessarily finite dimensional). We endow the group $\mathrm{AutTop}(\mathfrak{G})$ of all continuous automorphisms of $\mathfrak{G}$ with the compact-open topology.

THEOREM 7.11 (Burns-Spatzier). *The group* $\mathrm{AutTop}(\mathfrak{G})$ *is a locally compact metrizable group.*

PROOF. This follows from Burns-Spatzier [**18**] 2.1, in combination with Grundhöfer-Knarr-Kramer [**38**] 1.8 (there, it is proved that the metrizability assumption of Burns-Spatzier is superfluous), see also Bletz [**5**]. □

For example, the automorphism group of $\mathsf{H}_n(\mathbb{F},\mathbb{R})$ is a finite extension of the non-compact simple Lie group $\mathrm{PU}_{2,n-1}(\mathbb{F})$.

An *action* of a (Lie) group G on a compact quadrangle is a continuous homomorphism

$$G \longrightarrow \mathrm{AutTop}(\mathfrak{G}).$$

For example, the compact Lie group $\mathrm{U}_2(\mathbb{F})\times\mathrm{U}_{n-1}(\mathbb{F})$ acts transitively on the points, lines and flags of $\mathsf{H}_n(\mathbb{F},\mathbb{R})$. To get further, we need some facts about (compact) transformation groups.

7.C. Some results about compact transformation groups

We collect a couple of results which will be used in our classification of point homogeneous quadrangles. Let G be a group acting on a set X. We denote the image of G in the symmetric group of X by $G|_X$, and we put $G_X = \{g \in G |\ g = \mathrm{id}_X\}$. In other words, G_X is the *kernel of the action*, $G|_X$ is the *induced group*, and the sequence

$$1 \longrightarrow G_X \longrightarrow G \longrightarrow G|_X \longrightarrow 1$$

is exact. The G-action on X is effective if and only if $G_X = 1$. We put

$$\mathrm{Fix}(G,X) = \{x \in X |\ G\cdot x = \{x\}\}.$$

We will use Szenthe's solution of Hilbert's 5th problem:

THEOREM 7.12 (Szenthe). *Let G be a locally compact second countable group. Suppose that G acts effectively and transitively on a locally compact, locally contractible space X. Then G is a Lie group and $X \cong G/G_x$ is a manifold.*

PROOF. This is what is *proved* in Szenthe [**96**] (there, a more general result is claimed), cp. the remarks in Salzmann *et al.* 96.14. and in Grundhöfer-Knarr-Kramer [**38**] Thm. 2.2. □

PROPOSITION 7.13. *Let G be a compact connected Lie group acting transitively and effectively on a manifold X. Then*

$$\dim(G) \leq \binom{\dim(X)+1}{2}.$$

If equality holds, then $G = \mathrm{SO}(n+1)$ and $X = \mathbb{S}^n$ or $G = \mathrm{PSO}(n+1)$ and $X = \mathbb{RP}^n$.

PROOF. The idea of the proof is to introduce a G-invariant Riemannian metric on X. The stabilizer G_x of $x \in X$ acts effectively on T_xX, since every point $y \in X$ can be joined by a geodesic with x. Thus $G_x \subseteq \mathrm{O}(T_xX)$, whence $\dim(G_x) \leq \dim(\mathrm{O}(T_x)) = \binom{\dim(X)}{2}$, see Montgomery-Zippin [**74**] Thm. 6.2.5 and the following corollary, and Kobayashi-Nomizu [**53**] Vol.1 Note 10, Thm. 1, p. 308. □

The *type* of a G-orbit $G\cdot x$ in a set X is the conjugacy class of the stabilizer G_x. If G is a compact Lie group, then an orbit $G\cdot x$ is *principal* if the following two conditions hold.

PO$_1$ For every $y \in X$, the stabilizer G_y is conjugate to an overgroup of G_x (i.e. there exists an element $g \in G$ such that $G_{g(x)}$ fixes y).

PO$_2$ The set of all orbits of the same type as $G \cdot x$ is open and dense in X.

It is not difficult to prove that principal orbits exist, provided that G acts smoothly on a (smooth) manifold, see Bredon [**16**] IV Thm. 3.1, tom Dieck [**23**] Thm. 5.14. However, we need the general result which is essentially due to Montgomery-Yang [**73**] , and which is stated and proved in the present form in Biller [**4**] Thm. 2.2.3.

THEOREM 7.14. *Let G be a compact Lie group acting effectively on a connected integral ENR-manifold X. Then there exist principal orbits. Moreover, G acts effectively on each principal orbit.*

PROOF. For $x \in X$ let $d(x) = \dim(G \cdot x)$ and $c(x) = |\pi_0(G_x)|$. Let $X_r = \{x \in X |\ d(x) = r\}$ and $X_{r,v} = \{x \in X_r |\ c(x) = v\}$. Choose k as large as possible, such that $X_k \neq \varnothing$, and let $u = \min\{c(x) |\ x \in X_k\}$. Let $Y = X_{k,u}$. The existence of slices implies that Y is open, see Borel *et al.* [**11**] VIII Cor. 3.10. By Montgomery-Yang [**73**] Lemma 2, Y is dense in X, cp. Borel *et al.* [**11**] IX Lem. 3.2. Moreover, Y/G is connected by Borel *et al.* [**11**] IX Lem. 3.4. This together with the existence of slices readily implies that $Y \longrightarrow Y/G$ is a locally trivial fibre bundle, and all stabilizers of points in Y are conjugate.

If $g \in G$ fixes all points in some principal orbit, then clearly g fixes Y elemen-twise, and thus it fixes X, because Y is dense. Therefore, the action of G on each principal orbit is effective. □

COROLLARY 7.15. *Let G be a compact connected Lie group acting effectively on a connected n-dimensional integral ENR manifold X. Then*

$$\dim(G) \leq \binom{n+1}{2}.$$

If equality holds, then G acts transitively, and $X = \mathbb{S}^n$ or $X = \mathbb{RP}^n$, cp. Proposition 7.13 above. If the G-action is not transitive, then

$$\dim(G) \leq \binom{n}{2}.$$

PROOF. Let $Z \subseteq X$ be a principal G-orbit. By Proposition 7.13, G acts effectively on the compact connected manifold Z, so $\dim(G) \leq \binom{\dim(Z)+1}{2}$. If $Z \neq X$, then $\dim(Z) < \dim(X)$, since otherwise Z would be open in X. □

The inequalities can be improved using Mann's result [**67**]; further results are obtained in Biller's thesis [**4**].

If every point is contained in a principal orbit, then there is the following useful result due to Borel.

PROPOSITION 7.16 (Borel). *Let G be a compact Lie group acting on a space X. Suppose that all G-orbits have the same type, i.e. that all stabilizers G_x are conjugate to some group $H \subseteq G$. The set $\mathcal{H}$ of all G-conjugates of H can be identified with the homogeneous space $G/\mathrm{Nor}_G(H)$; in this way, it becomes a compact manifold. Then the map*

$$X \longrightarrow \mathcal{H}, \quad x \longmapsto G_x$$

is a continuous surjection (in fact a fibre bundle with $\mathrm{Fix}(H,X)$ *as typical fibre).*

PROOF. This is proved in Bredon [**16**] II.5.9. □

Let p be a prime, let $\mathbb{F}_p$ denote the field with p elements, and let A be an elementary abelian p-group of rank r, i.e. $A \cong (\mathbb{Z}/p)^r$.

THEOREM 7.17 (Floyd and Borel). *Let X a compact connected n-dimensional integral ENR manifold with the same integral homology as $\mathbb{S}^n$. Suppose that A acts effectively on X, and let $F = \mathrm{Fix}(A, X)$ denote the fixed point set of A. Then F has the same (sheaf theoretic) $\mathbb{F}_p$-cohomology as a sphere $\mathbb{S}^k$, for $-1 \leq k \leq n$ (where $\mathbb{S}^{-1} = \varnothing$). Note that $k < n$ if $r \geq 1$.*

If p is odd, then $n - k$ is even.

If $r \geq 1$, then there exists a subgroup $A' \subseteq A$ of index p such that $\mathrm{Fix}(A', X)$ is a $\mathbb{F}_p$-cohomology k'-sphere, for $k' > k$

PROOF. The first and second claim is proved in Borel *et al.* [**11**] IV 4.3, 4.4, 4.5. To prove the last claim we use the following result due to Borel. Let $\mathcal{A}$ denote the set of all subgroups of A of index p (so $|\mathcal{A}| = \frac{p^r-1}{p-1}$). For each $H \in \mathcal{A}$, let n_H denote the dimension of the cohomology sphere fixed by H. Then

$$n - k = \sum_{H \in \mathcal{A}} (n_H - k),$$

cp. Borel *et al.* [**11**] XIII Thm. 2.3. Put $k' = \max\{n_H | \ H \in \mathcal{A}\}$. Thus $(n - k)(p - 1) \leq (p^r - 1)(k' - k)$, and therefore $k' > k$ (note that $k < n$ if $r \geq 1$). □

7.D. Group actions on compact quadrangles

THEOREM 7.18 (Transitive actions on compact quadrangles).
Let $\mathfrak{G}$ be a compact connected quadrangle. Suppose that the topological automorphism group $\mathrm{AutTop}(\mathfrak{G})$ acts transitively on the point space $\mathcal{P}$. Then the following hold.

(i) *The group $\mathrm{AutTop}(\mathfrak{G})$ is a Lie group.*
(ii) *The quadrangle $\mathfrak{G}$ has finite dimension; in particular, the topological parameters (m_1, m_2) of $\mathfrak{G}$ are defined, and 7.10 applies to $\mathfrak{G}$.*
(iii) *If $m_1 \geq 2$, then there exists a compact connected Lie subgroup $G \subseteq \mathrm{AutTop}(\mathfrak{G})$ which acts transitively on $\mathcal{P}$.*

PROOF. The first claim follows from Szenthe's solution of Hilbert's 5th problem, see Theorem 7.12 above. Being a homogeneous space, $\mathcal{P}$ is a manifold and thus finite dimensional. The last claim follows from Montgomery [**72**] Cor. 3, because $\mathcal{P}$ is 1-connected. □

The following result is proved in Kramer [**54**].

THEOREM 7.19. *Let $\mathfrak{G}$ be a compact connected (m_1, m_2)-quadrangle and assume that the topological automorphism group acts transitively on the points or lines, as in 7.18. If $m_1 = m_2$, then $\mathfrak{G}$ is the real or complex symplectic quadrangle.*

PROOF. See Kramer [**54**] Thm. 5.2.4 and Thm. 5.2.3. □

If $\mathfrak{G}$ is a compact connected (m_1, m_2)-quadrangle with $m_1 \neq m_2$, then the cohomology of the point space is as in 7.10.

THEOREM 7.20. *Let $\mathfrak{G}$ be a compact connected (m_1, m_2)-quadrangle and assume that the topological automorphism group acts transitively on the points, as in 7.18. If $m_1 \neq m_2$, and if $m_1 \geq 3$, then there exists a compact connected Lie group $G \subseteq \mathrm{AutTop}(\mathfrak{G})$ such that (G, G_p) is one of the pairs in Theorem 3.15.*

PROOF. Since $m_1 \geq 3$, the point space $\mathcal{P}$ is 1-connected. By Theorem 7.18, there exists a compact Lie group G contained in the automorphism group which acts transitively on $\mathcal{P}$. Moreover, 7.10 shows that the homogeneous G-space $\mathcal{P}$ has the type of cohomology ring which we have classified in Theorem 3.15. □

The following lemma shows that several of the transitive actions occurring in Theorem 3.15 are not related to compact quadrangles.

LEMMA 7.21. *Let G be a compact Lie group acting effectively on a compact connected (m_1, m_2)-quadrangle $\mathfrak{G} = (\mathcal{P}, \mathcal{L}, \mathcal{F})$. Suppose that G acts transitively on $\mathcal{P}$. Let $N \subseteq G_p$ be a normal almost simple subgroup, and assume that G_p has no other almost simple subgroup which is of the same type as N. Then*

$$\dim(N) \leq \max\left\{\binom{m_1+1}{2}, \binom{m_2+1}{2}\right\}.$$

PROOF. It follows from the assumptions that N satisfies the condition (ii) of 7.3. Therefore N acts non-trivially on $D_1(p)$ or on $D_1(\ell)$. Since N is almost simple, this non-trivial action is necessarily almost effective (the kernel is finite). The claim follows now from 7.15. □

The following results are due to Biller [**4**] Sec. 5.1. We assume the following. $\mathfrak{G}$ is a compact connected (m_1, m_2)-quadrangle, A is an elementary abelian p-group, for some odd prime p, and of rank $r \geq 1$, acting effectively on $\mathfrak{G}$ and fixing a point row $D_1(\ell)$ pointwise.

LEMMA 7.22. *If m_2 is odd, then the fixed points and lines of A form a full, compact, and proper subquadrangle $\mathfrak{G}'$.*

PROOF. Let h be a line which is fixed pointwise by A, and let $q \in D_1(h)$. Then A fixes $h \in D_1(q)$, and in particular $\mathrm{Fix}(A, D_1(q))$ is non-empty. By Theorem 7.17, it is a cohomology sphere of positive dimension. Iterating this argument, we see that the substructure of $\mathfrak{Q}$ which is fixed by A is a full subquadrangle $\mathfrak{G}'$ (here we use the structure theorem for weak quadrangles as in Van Maldeghem [**110**] Thm. 1.6.2: the fixed structure is a weak subquadrangle with at least two thick lines and infinitely many thick points, so it is thick); since $\mathfrak{G}$ is compact, the set of all lines and points fixed by A is compact, so $\mathfrak{G}'$ is compact. Moreover, $\mathrm{Fix}(A, \mathcal{P}) \neq \mathcal{P}$, since we assumed that $A \neq 1$ acts effectively. □

Let $(m_1, m_2^{(r)})$ denote the parameters of $\mathfrak{G}' = \mathfrak{G}^{(r)} \subset \mathfrak{G} = \mathfrak{G}^{(0)}$. By Theorem 7.17, we find a subgroup $A_1 \subseteq A_0 = A$ of index p such that $\mathrm{Fix}(A_1, D_1(p))$ is strictly bigger than $\mathrm{Fix}(A_0, D_1(p))$, for a point $p \in \mathrm{Fix}(A, \mathcal{P})$. If we iterate this process, we obtain a sequence of full, compact, and proper subquadrangles $\mathfrak{G}^{(r)} \subset \mathfrak{G}^{(r-1)} \subset \cdots \subset \mathfrak{G}^{(0)} = \mathfrak{G}$, with parameters $(m_1, m_2^{(k)})$ for $\mathfrak{G}^{(k)}$, and $m_2^{(r)} < m_2^{(r-1)} < \cdots <$

$m_2^{(0)}$. By Proposition 7.9 we have the inequalities

$$\begin{aligned} m_1 &\le m_2^{(0)} - m_2^{(1)} \\ m_1 &\le m_2^{(1)} - m_2^{(2)} \\ m_1 &\le m_2^{(2)} - m_2^{(3)} \\ &\vdots \\ m_1 &\le m_2^{(r-1)} - m_2^{(r)} \end{aligned}$$

Adding these up, we have $rm_1 + m_2^{(r)} \le m_2^{(0)}$. In particular, $rm_1 < m_2$ holds for the parameters of $\mathfrak{G}$.

THEOREM 7.23 (Biller). *Let $\mathfrak{G}$ be a compact connected (m_1, m_2)-quadrangle, with m_2 odd, and let p be an odd prime. Let ℓ be a line, and let $(\mathbb{Z}/p)^r \cong A \subseteq$ $\mathrm{AutTop}(\mathfrak{G})$, be an elementary abelian p-group, for some odd prime p, and for some $r \ge 1$. Suppose that A fixes the point row $D_1(\ell)$ pointwise. Then $m_1 r < m_2$.*

PROOF. This follows from the previous discussion. □

COROLLARY 7.24 (Biller). *Let K be a compact connected Lie group acting effectively on a compact connected (m_1, m_2)-quadrangle and fixing a point row $D_1(\ell)$ pointwise. Assume that m_2 is odd. Then $\mathrm{rk}(K) < \frac{m_2}{m_1}$.*
In particular, if $m_1 = 4$ and $m_2 = 4k - 5$, then $\mathrm{rk}(K) \le k - 2$.

PROOF. We choose a maximal torus $T \subseteq K$, of rank, say, s. Thus we have $(\mathbb{Z}/3)^s \subseteq K$. □

7.E. The Stiefel manifolds

Put $G(n) = \mathrm{SO}(n), \mathrm{SU}(n), \mathrm{Sp}(n)$, let $\mathbb{F} = \mathbb{R}, \mathbb{C}, \mathbb{H}$ denote the corresponding skew field, and put $d = \dim_{\mathbb{R}} \mathbb{F} = 1, 2, 4$. In this section we classify those compact quadrangles which admit a line transitive $G(n)$-action, such that $\mathcal{L} = G(n)/G(n-2)$ is a Stiefel manifold, with $n \ge 5$ for $\mathbb{F} = \mathbb{R}, \mathbb{C}$ and $n \ge 4$ for $\mathbb{F} = \mathbb{H}$. Such a quadrangle has topological parameters

$$(d, d(n-1) - 1),$$

where $d = 1, 2, 4$ for $G = \mathrm{SO}(n), \mathrm{SU}(n), \mathrm{Sp}(n)$, respectively.

7.25. EXAMPLE Consider the quadrangle $\mathsf{H}_{n+1}(\mathbb{F}, \mathbb{R}) = (\mathcal{P}, \mathcal{L}, \mathcal{F})$ as in 7.8. Let $V \in \mathcal{L}$ be a 2-dimensional totally isotropic subspace of $\mathbb{F}^{n+2}$. It is not difficult to see that V admits a (unique) basis $\{u, v\}$ such that

$$\sum_{k=2}^{n+1} \bar{u}_k v_k = 0 \quad \text{and} \quad u_0 = v_1 = 1, \quad u_1 = v_0 = 0.$$

Thus $((u_2, \dots, u_{n+1}), (v_2, \dots, v_{n+1}))$ is an element of the Stiefel manifold $V_2(\mathbb{F}^n)$. This correspondence is in fact a $G(2) \times G(n)$-equivariant homeomorphism $\mathcal{L} \cong V_2(\mathbb{F}^n)$, and we may identify the line space $\mathcal{L}$ of $\mathsf{H}_{n+1}(\mathbb{F}, \mathbb{R})$ with $V_2(\mathbb{F}^n)$.

Next we note the following. If $k \ge 3$, then $\dim(G(k)) > \binom{d+1}{2}$, hence the almost simple group $G(k)$ cannot act non-trivially on a d-dimensional point row. This implies that $G(n)_\ell = G(n)_{p,\ell}$ for all points $p \in D_1(\ell)$, provided that $n \ge 5$. Thus, $G(n)_p \supseteq G_\ell = G(n-2)$. If $d = 4$ and $n = 4$, then $G(2) = \mathrm{Sp}(2) \cong \mathrm{Spin}(5)$ could act non-trivially. However, in this case the action would necessarily be transitive (we will see that this situation does not occur).

LEMMA 7.26. *Let $(\mathcal{P},\mathcal{L},\mathcal{F})$ be a compact generalized quadrangle, and assume that $G(n)$ acts transitive on the line space $\mathcal{L}$, such that there is a $G(n)$-equivariant homeomorphism $\mathcal{L} \cong G(n)/G(n-2) = V_2(\mathbb{F}^n)$. If $n \geq 5$, then the line stabilizer $G(n)_\ell = G(n-2)$ acts trivially on the point row $D_1(\ell)$, whence*

$$G(n)_p \supseteq G(n-2) = G(n)_\ell$$

for every $p \in D_1(\ell)$. The same conclusion holds for $n = 4$ and $\mathbb{F} = \mathbb{H}$.

PROOF. Only the last claim has to be proved. Thus we assume that $n = 4$ and $\mathbb{F} = \mathbb{H}$. If Sp(2) acts non-trivially on $D_1(\ell)$, then it acts transitively, because Sp(2) contains no subgroup of codimension less than 4 (this follows from 7.15). Thus Sp(4) acts transitively on the flags, and in particular transitively on the points. But we know from our classification 5.11 that Sp(4) cannot act transitively on a 1-connected space with the rational cohomology of $\mathbb{S}^4 \times \mathbb{S}^{15}$. □

We need to know the groups between $G(n-2)$ and $G(n)$.

LEMMA 7.27. *Let $H \subseteq G(n)$ be a connected almost simple subgroup such that $G(n-2) \subsetneq H \subsetneq G(n)$. Then H is conjugate to $G(n-1)$, provided that $n \geq 5$ for $\mathbb{F} = \mathbb{R}, \mathbb{C}$, and $n \geq 4$ for $\mathbb{F} = \mathbb{H}$.*

PROOF. Consider the adjoint representation of $G(n-2)$ on the Lie algebra $\mathfrak{g}(n)$ of $G(n)$. We decompose $\mathfrak{g}(n)$ into irreducible $G(n-2)$-modules. Then

$$\mathfrak{g}(n) = \mathfrak{g}(n-2) \oplus V \oplus V \oplus T,$$

where V is the natural $G(n-2)$-module, and $T \cong \mathbb{R}^{3d-2}$ is a trivial $G(n-2)$-module. Since $G(n-2) \subseteq H$, the Lie algebra $\mathfrak{h}$ of H is invariant under $G(n-2)$.

If $\mathfrak{h} \subseteq \mathfrak{g}(n-2) \oplus T$, then either $\mathfrak{h} = \mathfrak{g}(n-2)$ or $\mathfrak{h}$ is not almost simple; in either case, we get a contradiction to our assumptions. Thus there exists an element $(0, v_1, v_2, t) \in \mathfrak{h}$ with $\{0\} \neq \{v_1, v_2\}$. We consider the $G(n-2)$-orbit of this element.

If v_1, v_2 are $\mathbb{F}$-linearly independent, then the $G(n-2)$-orbit of $(0, v_1, v_2, t)$ is a (distorted) Stiefel manifold $V_2(\mathbb{F}^{n-2}) \cong G(n-2)/G(n-4)$ (here we use the assumption on n); thus, its $\mathbb{R}$-span contains $V \oplus V$. It follows that $V \oplus V \subseteq \mathfrak{h}$, and, consequently, that $\mathfrak{h} = \mathfrak{g}(n)$ and $H = G(n)$, contradicting our assumptions on H.

If $\{v_1, v_2\} \neq \{0\}$ are $\mathbb{F}$-linearly dependent, then the orbit of $(0, v_1, v_2, t)$ is a $(d(n-2)-1)$-sphere, and thus the linear span of the orbit contains an $(n-2)$-dimensional subspace W of $V \oplus V$ which is $G(n-2)$-isomorphic to V. The Lie algebra $\mathfrak{f}$ generated by W and $\mathfrak{g}(n-2)$ is isomorphic and conjugate to $\mathfrak{g}(n-1)$. Thus, $\mathfrak{f} \subseteq \mathfrak{h}$, and we have to prove that equality holds.

Assume that $(0, v_1', v_2', t') \in \mathfrak{h} \setminus \mathfrak{f}$. If $\{v_1', v_2'\} \neq \{0\}$, then one sees readily that $V \oplus V \subseteq \mathfrak{h}$ and hence $\mathfrak{h} = \mathfrak{g}(n)$, again a contradiction to out assumptions.

We are left with the case that $\mathfrak{h} = \mathfrak{g}(n-2) \oplus W \oplus T'$, where $W \subseteq V \oplus V$ and $T' \subseteq T$. We may assume that $\mathfrak{f} = \mathfrak{g}(n-1)$ and that $W = V \oplus 0$. Thus $\mathfrak{h} \subseteq \mathfrak{g}(n-1) \oplus T''$, where T'' is of dimension $d-1$. Since we assumed that $\mathfrak{h}$ is simple, this implies that $\mathfrak{h} \cong \mathfrak{g}(n-1)$. □

Now we go back to the quadrangle. We assume that $n \geq 5$ for $\mathbb{F} = \mathbb{R}, \mathbb{C}$ and $n \geq 4$ for $\mathbb{F} = \mathbb{H}$. Then $G(n)_{p,\ell} = G(n)_\ell = G(n-2)$ for all $p \in D_1(\ell)$, and we have inequalities

$$\begin{aligned}
\dim(G(n)_p) &\geq \dim(G(n)) - \dim(\mathcal{P}) \\
\dim(G(n)_\ell) &= \dim(G(n)) - \dim(\mathcal{L}) \\
\dim(D_1(p)) &\geq \dim(G(n)_p) - \dim(G(n)_\ell)
\end{aligned}$$

whence

$$d(n-2) - 1 \leq \dim(G(n)_p) - \dim(G(n)_\ell) \leq d(n-1) - 1.$$

Put $(G(n)_p)^\circ = K_1 \cdot K_2$, where K_1 is almost simple, $G(n-2) = G(n)_\ell \subseteq K_1$, and $K_2 \subseteq \mathrm{Cen}_{G(n)}(G(n-2))$. Note that $\dim(\mathrm{Cen}_{G(n)}(G(n-2)) = 3d - 2$. If $\dim(G(n)_p) - \dim(G(n)_\ell) \geq \dim(\mathrm{Cen}_{G(n)}(G(n-2))$, then $K_1 \neq G(n-2)$ by the inequalities above. It follows then from 7.27 that K_1 is conjugate to $G(n-1)$. Thus $\dim(K_1/G(n-2)) = d(n-1) - 1$. This is the full dimension of a line pencil. Therefore G_p acts transitively on $D_1(p)$, and thus $G_p = K_1$ is connected and conjugate to $G(n-1)$.

Now $d(n-2) - 1 \geq 3d - 2$ for $d = 1, 2, 4$, provided that $n \geq 5$. In these cases, we have proved that $G(n)_p$ is conjugate to $G(n-1)$, contains $G(n-2)$ and acts transitively on the line pencil $D_1(p)$.

Finally, assume that $n = 4$ and $\mathbb{F} = \mathbb{H}$. Then $21 \leq \dim(G_p) \geq \dim(\mathrm{Sp}(4)) - \dim(\mathcal{P}) = 36 - 19 = 17$. In fact, $\mathrm{Sp}(4)$ is not transitive on $\mathcal{P}$ by 5.11, so we have strict inequality $\dim(G_p) > 17$. Assume that $K_1 = G(n-2)$. Then $\dim(K_2) \geq 8$, and $K_1 \subseteq \mathrm{Cen}_{\mathrm{Sp}(4)}(Sp(2))^\circ \cong \mathrm{Sp}(2)$. But $\mathrm{Sp}(2)$ has no subgroup of codimension 2 (by 7.15). Thus K_1 is strictly bigger than $G(n-2)$, and thus $K_1 \cong G(n-1) = G(n)_p$.

THEOREM 7.28. *Let $G(n)$ denote one of the groups $\mathrm{SO}(n), \mathrm{SU}(n), \mathrm{Sp}(n)$. Suppose that $G(n)$ acts line transitively on a compact quadrangle $\mathfrak{G}$, such that $\mathcal{L} \cong G(n)/G(n-2)$. If $\mathbb{F} = \mathbb{R}, \mathbb{C}$ and $n \geq 5$, or if $\mathbb{F} = \mathbb{H}$ and $n \geq 4$, then $\mathfrak{G}$ is uniquely determined and continuously G-isomorphic to the classical quadrangle $\mathsf{H}_{n+1}\mathbb{F}$.*

PROOF. The discussion above shows that $G(n)_p$ acts transitively on $D_1(p)$ for all $p \in \mathcal{P}$. Put $\mathcal{G} = \{G(n)_p |\, p \in D_1(\ell)\}$. The quadrangle is uniquely determined by the triple

$$(G(n), G(n-2), \mathcal{G}),$$

cp. 7.2. It remains to determine $\mathcal{G}$ in group theoretic terms. Let $\mathcal{G}'$ denote the set of all conjugates of $G(n-1)$ which contain $G(n-2)$. Clearly, $\mathcal{G} \subseteq \mathcal{G}'$; note that $\mathcal{G}'$ is the collection of subgroups which arises from the corresponding classical quadrangle. We wish to show that $\mathcal{G} = \mathcal{G}'$.

The map $p \longmapsto G(n)_p$ is continuous by 7.16. If $p, q \in D_1(\ell)$ are distinct points, then $G(n)_p \neq G(n)_q$, since otherwise $G(n)_p$ would fix $\ell \in D_1(p)$. Thus, the map $q \longmapsto G(n)_q$ is injective on the point row $D_1(\ell)$, and its image is $\mathcal{G}$. Therefore $\mathcal{G}$ is a homology d-sphere inside $\mathcal{G}'$. From the corresponding classical quadrangle we see that $\mathcal{G}'$ is in fact homeomorphic to $\mathbb{S}^d$. It follows that $\mathcal{G} = \mathcal{G}'$.

We have established an abstract isomorphism between $\mathfrak{G}$ and a certain classical quadrangle,

$$\mathfrak{G} \xrightarrow{\ \phi\ } \mathsf{H}_{n+1}(\mathbb{F}, \mathbb{R}).$$

The map ϕ is G-equivariant and continuous on the line space $\mathcal{L}$, so it is continuous on every line pencil. Every point row $D_1(\ell)$ can be embedded in $\mathcal{L}$ by the map $q \longmapsto \mathrm{proj}_q h$, where $h \in D_4(\ell)$ is some line. Thus, ϕ is continuous on every point row and every line pencil. By Bödi-Kramer [**7**] Prop. 3.5, the map ϕ is continuous (and so is its inverse). □

The last step in the proof will be needed again later, so we state it separately.

PROPOSITION 7.29. *Let* $\mathfrak{G} \xrightarrow{\phi} \mathfrak{G}'$ *be an abstract isomorphism of generalized quadrangles. If* $\mathfrak{G}$ *and* $\mathfrak{G}'$ *are topological quadrangles, and if* ϕ *is continuous on the line or point set, then* ϕ *is continuous everywhere.*

PROOF. The proof is the same as the one given above. □

7.F. The $(4, 4n-5)$-series

Here $G/G_p = \mathrm{Sp}(n) \times \mathrm{Sp}(2)/\mathrm{Sp}(n-1)\cdot\mathrm{Sp}(1)\cdot\mathrm{Sp}(1)$. The canonical examples for such an action are the quadrangles $\mathsf{H}_{n+1}(\mathbb{H},\mathbb{R})$.

7.30. EXAMPLE We use the same notation as in 7.8. Every 1-dimensional totally isotropic subspace in $\mathbb{H}^{n+2}$ is spanned by a vector $(u_0, u_1, \dots, u_n, u_{n+1})$, with $|u_0|^2 + |u_1|^2 = 1$. This vector is not unique, but the pair of vectors

$$(u_2\bar{u}_0, \dots, u_{n+1}\bar{u}_0, u_2\bar{u}_1, \dots, u_{n+1}\bar{u}_1)$$

in $\mathbb{H}^{2\times n}$ is. This embedding of the point space $\mathcal{P}$ in $\mathbb{H}^{2\times n}$ is $\mathrm{Sp}(2)\times\mathrm{Sp}(n)$-equivariant. The image is a focal manifold of the homogeneous isoparametric foliation with $g = 4$ and $(m_1, m_2) = (4, 4n-5)$. The other focal manifold is the image of the line space $\mathcal{L}$ considered in 7.25 before. *Mutatis mutandis,* similar remarks apply to $\mathbb{F} = \mathbb{R}, \mathbb{C}$ and also to $\mathbb{F} = \mathbb{O}$, see Kramer [**57**] [**58**] [**55**].

This example gives us a linear model for this homogeneous space, i.e. an equivariant embedding into a vector space. We let $V = \mathbb{H}^n$ denote the natural $\mathrm{Sp}(n)$-module and consider $V \oplus V$. Then

$$\mathrm{Cen}_{\mathrm{SO}(8n)}(\mathrm{Sp}(n))^\circ \cong \mathrm{Sp}(2);$$

consequently, we have an (irreducible) action of $\mathrm{Sp}(n)\times\mathrm{Sp}(2)$ on $V \oplus V$. We can view $V\oplus V$ as the set of all $2\times n$-matrices with entries in $\mathbb{H}$; the group $\mathrm{Sp}(n)$ acts by left multiplication, and $\mathrm{Sp}(2)$ acts by conjugate transpose right multiplication. The orbit of the vector $p = (e_1, 0)$ is precisely the set

$$\mathcal{P} = \{(xc, xs) \in V \oplus V \mid x \in V, |x| = 1, c, s \in \mathbb{H}, |c|^2 + |s|^2 = 1\},$$

and the stabilizer of p is of the form

$$\left\{ \begin{pmatrix} a & \\ & A \end{pmatrix} \times \begin{pmatrix} a & \\ & b \end{pmatrix} \middle| \; A \in \mathrm{Sp}(n-1), a, b \in \mathbb{S}^3 \subseteq \mathbb{H} \right\}.$$

Thus $\mathcal{P} \cong \mathrm{Sp}(n)\times\mathrm{Sp}(2)/\mathrm{Sp}(n-1)\cdot\mathrm{Sp}(1)\cdot\mathrm{Sp}(1)$. Consider the map

$$\phi : \mathcal{P} \longrightarrow \mathbb{S}^4 \subseteq \mathbb{R}\oplus\mathbb{H}, \quad (xc, xs) \longmapsto (|c|^2 - |s|^2, 2\bar{c}s).$$

The group $G = \mathrm{Sp}(n)\times\mathrm{Sp}(2)$ acts — via the second factor — as $\mathrm{SO}(5)$ on $\mathbb{R}\oplus\mathbb{H}$, and the map ϕ is G-equivariant.

Our aim is to show that G acts transitively on the flag space $\mathcal{F}$ and hence on the line space $\mathcal{L}$. Let $G_{[\ell]}$ denote the kernel of the G_ℓ-action on the point row $D_1(\ell)$,

$$1 \longrightarrow G_{[\ell]} \longrightarrow G_\ell \longrightarrow G_\ell|_{D_1(\ell)} \longrightarrow 1.$$

Let $H = \mathrm{Sp}(n) \subset G$ denote the normal subgroup acting from the left. Every H-orbit in $\mathcal{P}$ is homeomorphic to a sphere $\mathbb{S}^{4n-1}$; these orbits are precisely the fibres of ϕ (these fibres turn out to be ovoids, cp. Kramer-Van Maldeghem [**59**] and Kramer [**57**] [**58**]). Let p be a point. Then $\mathrm{Fix}(H_p, H \cdot p) \cong \mathbb{S}^3$, and $\mathrm{Fix}(H_p, \mathcal{P}) \cong \mathbb{S}^7$. Thus $p^\perp \not\subseteq \mathrm{Fix}(H_p, \mathcal{P})$ if $n \geq 3$. So assume that $n \geq 3$ and let $q \in D_2(p)$ be a point which is not fixed by H_p. Then $H_{p,q}$ is conjugate to $\mathrm{Sp}(n-2)$. The group $H_{p,q}$ is almost simple and of dimension at least 21; thus, it fixes the point row $D_1(\ell)$ pointwise by Proposition 7.15. Now $\mathrm{rk}(G_{[\ell]}) \geq \mathrm{rk}(H_{p,q}) = n-2$. On the other hand, $\mathrm{rk}(G_{[\ell]}) \leq n-2$ by Corollary 7.24. The group $H_{p,q}$ is almost simple and of the same rank as $(G_{[\ell]})^\circ$, so $(G_{[\ell]})^\circ$ itself is almost simple. If $n \geq 4$, then Lemma 7.27 shows that $(G_{[\ell]})^\circ \cong \mathrm{Sp}(n-2)$.

The dimension of $D_1(\ell)$ is 4, so $G_\ell|_{D_1(\ell)}$ is at most 10-dimensional by Proposition 7.15, and if the group is 10-dimensional, then it is locally isomorphic to $\mathrm{SO}(5)$ and transitive. We have

$$\begin{aligned}
\dim(\mathrm{Sp}(n-2)) + 10 &= \dim(G) - \dim(\mathcal{L}) \\
&\leq \dim(G_\ell) \\
&= \dim(G_\ell|_{D_1(\ell)}) + \dim(G_{[\ell]}) \\
&\leq \dim(G_{[\ell]}) + 10 \\
&= \dim(\mathrm{Sp}(n-2)) + 10.
\end{aligned}$$

Therefore G_ℓ acts transitively on $D_1(\ell)$, and G acts transitively on $\mathcal{L}$. By our previous classification, H acts transitively on $\mathcal{L}$, and $\mathfrak{G}$ is the standard hermitian quadrangle over the quaternions.

THEOREM 7.31. *Suppose that* $\mathrm{Sp}(n)\cdot\mathrm{Sp}(2)$ *acts point transitively on a compact quadrangle* $\mathfrak{G}$, *such that* $\mathcal{P} \cong \mathrm{Sp}(n)\cdot\mathrm{Sp}(2)/\mathrm{Sp}(n-1)\cdot\mathrm{Sp}(1)\cdot\mathrm{Sp}(1)$. *If* $n \geq 4$, *then* $\mathfrak{G}$ *is uniquely determined and continuously G-isomorphic to the classical quadrangle* $\mathsf{H}_{n+1}(\mathbb{H}, \mathbb{R})$. *Moreover, Sp(n)* $\cdot\,\mathrm{Sp}(2)$ *acts transitively on* $\mathcal{L}$ *and on* $\mathcal{F}$. □

7.G. Products of spheres

We consider the following situation: the compact connected Lie group G acts on a compact quadrangle in such a way that

$$\mathcal{P} = G/H = K_1/H_1 \times K_2/H_2 = \mathbb{S}^{m_1} \times \mathbb{S}^{m_1+m_2}$$

is a product of homogeneous spheres. We use the fact that the inclusion of a point row $D_1(\ell) \longrightarrow \mathcal{P}$ induces a map onto $\pi_{m_1}(\mathcal{P})$. The composite

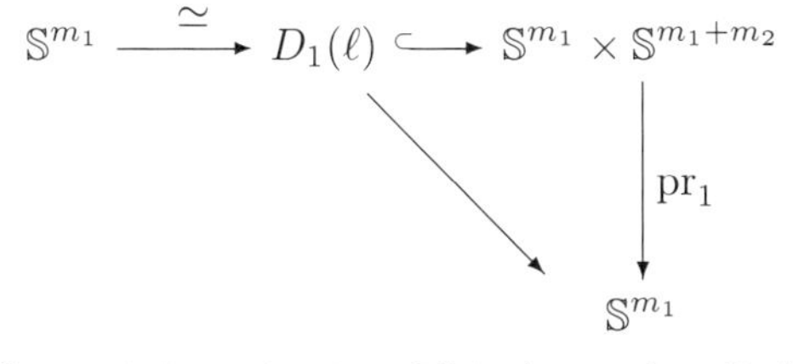

has non-zero degree, hence it is surjective. This shows that $D_1(\ell) \cap (\{x\} \times \mathbb{S}^{m_1+m_2}) \neq \varnothing$ for every $x \in \mathbb{S}^{m_1}$. The stabilizer H_1 fixes a set containing $\mathbb{S}^0 \times \mathbb{S}^{m_1+m_2}$. But every point row $D_1(\ell)$ meets this set twice. Therefore H_1 fixes every point row, and $(K_1, H_1) = (\mathrm{Sp}(1), 1)$. Thus $m_1 = 3$.

PROPOSITION 7.32. *Let G be a compact connected Lie group. If the point space $\mathcal{P}$ of a compact quadrangle is a product of homogeneous spheres*

$$\mathcal{P} = G/H = K_1/H_1 \times K_2/H_2 = \mathbb{S}^{m_1} \times \mathbb{S}^{m_1+m_2}$$

as above, then $m_1 = 3$, m_2 is even, and $(K_1, H_1) = (\mathrm{Sp}(1), 1)$. □

Now we use Lemma 7.21 to exclude some more cases. If H_2 is almost simple, then

$$\dim(H_2) \leq \max\left\{\binom{4}{2}, \binom{m_2+1}{2}\right\}.$$

This excludes the pairs $(K_2, H_2) = (\mathrm{SO}(m_2+4), \mathrm{SO}(m_2+3))$, for $m_2 \geq 2$, and $(K_2, H_2) = (\mathrm{Spin}(7), \mathrm{G}_2)$. We are left with the homogeneous spaces

$$\mathrm{Sp}(n)/\mathrm{Sp}(n-1) \times \mathrm{Sp}(1) \text{ for } n \geq 2$$
$$\mathrm{SU}(n+1)/\mathrm{SU}(n) \times \mathrm{Sp}(1) \text{ for } n \geq 2$$
$$\mathrm{Spin}(9)/\mathrm{Spin}(7) \times \mathrm{Sp}(1).$$

There exist point homogeneous compact connected quadrangles with the first type of group action, see Example 8.13 in Chapter 8.

7.H. Summary

We summarize the main results of this chapter.

THEOREM 7.33. *Let $\mathfrak{G} = (\mathcal{P}, \mathcal{L}, \mathcal{F})$ be a compact connected quadrangle. Suppose that the topological automorphism group* $\mathrm{AutTop}(\mathfrak{G})$ *acts transitively on the point space $\mathcal{P}$. Assume that the topological parameters (m_1, m_2) of $\mathfrak{G}$ satisfy the condition $m_1 \geq 2$. Then there exists a compact connected Lie group $G \subseteq \mathrm{AutTop}(\mathfrak{G})$ which acts transitively and irreducibly on $\mathcal{P}$.*

(A) *If $m_1 = m_2$, then $m_1 = m_2 = 2$, the group $G = \mathrm{SO}(5)$ acts transitively on the flag space $\mathcal{F}$ and the point space $\mathcal{P}$, and $\mathfrak{G}$ is G-equivariantly isomorphic to the complex symplectic quadrangle $W(\mathbb{C})$ or its dual, the complex orthogonal quadrangle $Q_4(\mathbb{C})$.*

(B) *If $m_1 \geq 3$, then $m_1 \neq m_2$. We have classified the following subcases.*

(B1) *If $(m_1, m_2) = (4, 4n-5)$, for $n \geq 4$, then $G = \mathrm{Sp}(n) \cdot \mathrm{Sp}(2)$ and $\mathfrak{G}$ is G-equivariantly isomorphic to the quaternion hermitian quadrangle $\mathsf{H}_{n+1}(\mathbb{H}, \mathbb{R})$.*

(B2) *If $(m_1, m_2) = (4n-5, 4)$, for $n \geq 4$, then $G = \mathrm{Sp}(n)$ and $\mathfrak{G}$ is G-equivariantly isomorphic to the dual of the quaternion hermitian quadrangle $\mathsf{H}_{n+1}(\mathbb{H}, \mathbb{R})$.*

(B3) *If $(m_1, m_2) = (2n-3, 2)$, for $n \geq 5$, then $G = \mathrm{SU}(n)$ and $\mathfrak{G}$ is G-equivariantly isomorphic to the dual of the complex hermitian quadrangle $\mathsf{H}_{n+1}(\mathbb{C}, \mathbb{R})$.*

(B4) *If $(m_1, m_2) = (n-1, 1)$, for $n \geq 9$, then $\mathfrak{G}$ is G-equivariantly isomorphic to the dual of the real orthogonal quadrangle $Q_{n+1}(\mathbb{R})$.*

(B5) *In the remaining cases $(G, \mathcal{P})$ is one of the pairs determined in Theorem 3.15 (B) and (C).*

PROOF. Case (A) was proved in Kramer [**54**]. The cases (B1)-B(4) were considered in the previous sections. □

Note that the theorem above does not summarize *all* situations which were considered in this chapter; we classified some more quadrangles.

COROLLARY 7.34. *Let $\mathfrak{G} = (\mathcal{P}, \mathcal{L}, \mathcal{F})$ be a compact connected quadrangle. Suppose that the topological automorphism group* $\mathrm{AutTop}(\mathfrak{G})$ *acts transitively on the point space $\mathcal{P}$. Assume that the topological parameters (m_1, m_2) of $\mathfrak{G}$ satisfy the condition*

$$m_1 \geq 10.$$

Then $\mathfrak{G}$ is G-equivariantly isomorphic to the dual of one of the classical compact connected Moufang quadrangles $\mathsf{Q}_n(\mathbb{R})$, $\mathsf{H}_n(\mathbb{C}, \mathbb{R})$, or $\mathsf{H}_n(\mathbb{H}, \mathbb{R})$.

PROOF. By 7.32, the point space $\mathcal{P}$ cannot be a product of homogeneous spheres. Thus $\mathcal{P}$ is a Stiefel manifold by 3.16. □

Case (B5) in the theorem above consists mainly of infinite series

$$\mathrm{Sp}(1) \times (\mathrm{Sp}(n)/\mathrm{Sp}(n-1))$$
$$\mathrm{Sp}(1) \times (\mathrm{SU}(n+1)/\mathrm{SU}(n))$$
$$\mathrm{Sp}(n) \times \mathrm{SU}(3)/\mathrm{Sp}(n-1) \cdot \mathrm{Sp}(1)$$
$$\mathrm{Sp}(n) \times \mathrm{Sp}(2)/\mathrm{Sp}(n-1) \cdot \mathrm{Sp}(1)$$

and a finite collection of sporadic spaces. I conjecture that only the first series corresponds to quadrangles. The following related result is proved by Biller [**4**]; it improves some of our results obtained in this chapter.

THEOREM 7.35 (Biller). *Let $\mathfrak{G}$ be a compact connected $(4, 4n-5)$-quadrangle, for $n \geq 2$, and assume that a compact connected Lie group G of dimension at least $\binom{2n+1}{2}+10$ acts effectively on $\mathfrak{G}$. Then $\mathfrak{G}$ is continuously isomorphic to $\mathsf{H}_{n+1}(\mathbb{H}, \mathbb{R})$, and $G \cong \mathrm{Sp}(2) \cdot \mathrm{Sp}(n)$ acts transitively on the flags.*

If a compact connected Lie group G of dimension at least 22 acts effectively on a compact connected $(4, 5)$-quadrangle $\mathfrak{G}$, then G acts transitively on the flags and $\mathfrak{G}$ is continuously isomorphic to a certain classical quadrangle (the so-called anti-hermitian quadrangle over the quaternions).

PROOF. See Biller [**4**] Sec. 5.3. □

Some more cases (e.g. $\mathrm{E}_6/\mathrm{F}_4$ or $\mathrm{SU}(6)/\mathrm{Sp}(3)$) can be excluded by 7.21. More cases can be excluded using the following deep result of Stolz and Markert.

THEOREM 7.36 (Markert). *Let $\mathfrak{G}$ be a compact connected (m_1, m_2)-quadrangle, with $2 \leq m_1 < m_2$. Put $k = \min\{m_2 - m_1, m_1 - 1\}$. Then*

$$2^{\phi(k)} \text{ divides } m_1 + m_2 + 1,$$

where $\phi(k) = |\{i|\ 1 \leq i \leq k \text{ and } i \equiv 0, 1, 2, 4 \pmod 8\}|$.

PROOF. This follows from results of Stolz [**91**]. However, the proof requires several non-trivial modifications of Stolz' paper, see Markert [**68**]. □

A complete classification of all compact connected quadrangles which admit a point or line transitive automorphism group seems to be difficult, but not impossible.

7.37. CONJECTURE *A compact connected quadrangle which admits a point or line transitive automorphism group is either a Moufang quadrangle or a quadrangle associated to an isoparametric hypersurface of Clifford type (with parameters $(3, 4k)$ or $(8, 7)$).*

CHAPTER 8

Homogeneous focal manifolds

We apply our classification of homogeneous spaces to obtain a classification of homogeneous focal manifolds of isoparametric hypersurfaces. A closed submanifold $M^n \subseteq \mathbb{R}^{n+k}$ is called isoparametric if its normal bundle is (globally) flat, and if the eigenvalues of the Weingarten map along any parallel normal field are constant. Typical examples are principal orbits of isotropy representations of Riemannian symmetric spaces. However, Ferus-Karcher-Münzner showed that there are infinitely many isoparametric hypersurfaces which do *not* arise from isotropy representations of symmetric spaces. On the other hand, Thorbergsson proved that an irreducible full isoparametric submanifold with codimension strictly bigger than 2 (i.e. M is not congruent to a product, the codimension is $k \geq 3$, and M is not contained in an affine hyperplane) is a principal orbit of the isotropy representation of a Riemannian symmetric space. This should be compared to Tits' result about buildings of higher rank mentioned in the previous chapter.

This correspondence is not accidental. The rank of an isoparametric submanifold $M^n \subseteq \mathbb{R}^{n+k}$ is defined as $\dim(\langle M\rangle_{\text{aff}})-\dim(M)$, where $\langle M\rangle_{\text{aff}}$ denotes the affine span of M, i.e. the intersection of all affine hyperplanes containing M. It is called irreducible if it is not congruent to a product $M_1 \times M_2 \subseteq \mathbb{R}^{m_1} \times \mathbb{R}^{m_2}$ of isoparametric submanifolds $M_1 \subseteq \mathbb{R}^{m_1}$, $M_2 \subseteq \mathbb{R}^{m_2}$ under an isometry of $\mathbb{R}^{m_1+m_2}$. Thorbergsson [**101**] proved that an irreducible isoparametric submanifold $M^n \subseteq \mathbb{R}^{n+k}$ of rank $k \geq 3$ carries in a natural way the structure of an irreducible spherical building of rank k. Using Tits' classification [**106**] (combined with a result by Burns-Spatzier [**18**]), he deduces that the building is Moufang and arises as the spherical building at infinity of a Riemannian symmetric space G/K. The isotropy representation of K on G/K yields an isoparametric family in $\mathfrak{p}$, where $\mathfrak{g} = \mathfrak{k} \oplus \mathfrak{p}$ is a Cartan decomposition of G. A closer inspection shows that there is an isometry $\mathbb{R}^{n+k} \cong \mathfrak{p}$ (up to some real scaling factor) which carries M onto an isoparametric submanifold in $\mathfrak{p}$ of the family belonging to G/K. Thus he obtains the following classification [**101**].

Theorem (Thorbergsson) *A closed irreducible isoparametric submanifold of rank at least 3 arises from the isotropy representation of an irreducible Riemannian symmetric space of non-compact type.*

An isoparametric submanifold of rank 2 is (up to some real scalar) congruent to an isoparametric hypersurface in the unit sphere of the ambient vector space. Ferus-Karcher-Münzner [**34**] constructed an infinite series of families of isoparametric hypersurfaces in spheres which are not congruent (not even homeomorphic) to principal orbits of isotropy representations of Riemannian symmetric spaces. So Thorbergsson's Theorem does not carry over to the rank 2 case. However, a weak form of his result (which is due to Hsiang-Lawson [**45**] and much older than Thorbergsson's result) is still valid in the rank 2 situation.

Theorem (Hsiang-Lawson) *A closed irreducible isoparametric submanifold of rank 2 arises from the isotropy representation of an irreducible Riemannian symmetric space of non-compact type and rank 2, provided that it admits a transitive isometry group.*

Table II in Hsiang-Lawson [**45**] is incomplete (the hypersurface with 6 distinct principal curvatures and multiplicities $(1,1)$ is erroneously omitted, cp. Uchida [**109**]); a corrected table can be found in Takagi-Takahashi [**98**].

Note also that in the case of higher rank Thorbergsson *proves* the existence of a transitive isometry group (the isotropy group K of the symmetric space G/K). Recently, Heintze-Liu [**43**] gave a new proof for the existence of a transitive isometry group in the case of rank at least 3; this yields also a new proof of Thorbergsson's Theorem. Yet another proof is due to Olmos [**79**]. Most of the new examples discovered by Ferus-Karcher-Münzner admit no transitive isometry groups.

Generalizing the result of Hsiang-Lawson [**45**], we consider the following situation: M is an irreducible isoparametric hypersurface whose isometry group acts transitively on one of the focal manifolds of M. Some (but not all) of the Ferus-Karcher-Münzner examples have this property. By Münzner [**77**], the spectrum of the Weingarten map (i.e. the set of all principal curvatures) of an isoparametric hypersurface has $g = 1, 2, 3, 4$ or 6 elements. The cases $g = 1, 2$ are easy to classify, so the interesting cases are $g = 3, 4, 6$. For $g = 3, 6$ one can show that a group which acts transitively on one focal manifold acts also transitively on the isoparametric hypersurface itself. Also, the case $g = 3$ was completely classified by Cartan, and the case $g = 6$ was partially classified by Dorfmeister-Neher (see Section 8.A below for references). Thus, the interesting case is $g = 4$. In this chapter we prove the following result.

Theorem A *Let M be a closed isoparametric hypersurface, with 4 distinct principal curvatures. Assume that the isometry group of M acts transitively on one of the two focal manifolds of M, and that this focal manifold is 2-connected. Then either M itself is homogeneous, or M is of Clifford type, with multiplicities $(8, 7)$ or $(3, 4k)$.*

Wolfrom [**114**] recently extended this result to the case where one multiplicity is $m_1 = 2$ (for all g), showing that no new examples occur. If $g = 4, 6$, and if one of the multiplicities is $m_1 = 1$, then the results of Takagi [**97**] and Dorfmeister-Neher [**28**] apply. Combining Theorem A with these results, we have the following theorem.

Theorem B *Let M be a closed irreducible isoparametric submanifold of rank at least 2. Assume that the isometry group of M acts transitively on one of the focal manifolds of M. Then either M itself is homogeneous and arises from the isotropy representation of an irreducible Riemannian symmetric space (of non-compact type), or M is of Clifford type, with multiplicities $(8, 7)$ or $(3, 4k)$.*

Finally, I would like to mention Immervoll's new and beautiful result [**48**].

Theorem (Immervoll) *Let M be a closed irreducible isoparametric submanifold of rank at least 2; if the rank is 2, assume we are not in the situation $(g, m_1, m_2) = (6, 2, 2)$. Then the simplicial complex associated to M is a compact connected Tits building.*

Probably, the case $(g, m_1, m_2) = (6, 2, 2)$ is not really an exception; it is conjectured that only one isoparametric hypersurface with these parameters exists (and this known hypersurface is a building, the so-called split Cayley hexagon over the complex numbers).

The strategy of the proof of Theorem A is as follows. In the situation of Theorem A, the focal manifold has the same integral cohomology as a product of spheres. Thus we can apply our classification of homogeneous spaces. However, not every homogeneous space in 3.15 is a focal manifold. The fact that the homogeneous space admits an equivariant embedding into a vector space of the right dimension rules out many of the 'wrong' manifolds. If the homogeneous space is one of the known examples of focal manifolds, then it is in most cases easy to see that this equivariant embedding is unique. Finally, some low-dimensional cases and some exceptional spaces need special attention.

Our method does not use very much Riemannian geometry; what we use is the global behavior of isoparametric submanifolds, and also the point-line geometry associated to such a submanifold. In a sense, we treat these submanifolds as if they were nicely embedded Tits buildings; many proofs are similar to the proofs in the last chapter. I think that both the effectiveness of this method and the close relation between the known isoparametric submanifolds and compact buildings justify this approach.

There are many papers and several books on isoparametric hypersurfaces and submanifolds, we just mention (in alphabetic order) Abresch [**2**], Cecil-Ryan [**21**], Console-Olmos [**22**], Dorfmeister-Neher [**25**], [**26**], [**27**], [**28**], Ferus-Karcher-Münzner [**34**], Grove-Halperin [**36**], Heintze-Liu [**43**], Hsiang-Palais-Terng [**46**], Knarr-Kramer [**52**], Münzner [**76**] [**77**], Olmos [**79**], Palais-Terng [**82**], Stolz [**91**], Strübing [**95**], Terng [**100**], Thorbergsson [**101**], [**102**], [**103**], Wang [**112**]. Thorbergsson's survey [**103**] is a good introduction to the subject and contains many references; a somewhat older bibliography is Kühnel-Cecil [**64**].

8.A. Isoparametric hypersurfaces

Let $X \subseteq \mathbb{S}^{r+1}$ be a submanifold. We denote the normal bundle of $X \subseteq \mathbb{S}^{r+1}$ by $\perp X$.

8.1. ISOPARAMETRIC HYPERSURFACES

A compact connected hypersurface $\mathcal{F} \subseteq \mathbb{S}^{r+1}$ is called *isoparametric* if its principal curvatures are constant. Fix a unit normal field N and consider the normal exponential map

$$\mathbb{R} \times \mathcal{F} \longrightarrow \mathbb{S}^{r+1}, \qquad (t,x) \longmapsto \exp_x(tN).$$

If t is small enough, then the endpoint map $\eta_t : x \longmapsto \exp_x(tN)$ is a diffeomorphism, and the image of this map is a parallel hypersurface which is also isoparametric. More precisely, there are real numbers $\theta_1 < 0 < \theta_2$ such that η_t is a diffeomorphism for $\theta_1 < t < \theta_2$; if $t = \theta_1, \theta_2$, then η_t is a submersion onto a manifold of strictly smaller dimension. Put

$$\mathcal{P} = \eta_{\theta_2}(\mathcal{F}) \text{ and } \mathcal{L} = \eta_{\theta_1}(\mathcal{F})$$

and

$$\mathrm{pr}_{\mathcal{P}} = \eta_{\theta_2} \text{ and } \mathrm{pr}_{\mathcal{L}} = \eta_{\theta_1}.$$

The parallel hypersurfaces of $\mathcal{F}$ are again isoparametric; thus, we have a foliation of the sphere by isoparametric hypersurfaces and the two focal manifolds, a so-called *isoparametric foliation.*

Let g denote the number of distinct principal curvatures of $\mathcal{F}$. A remarkable result of Münzner [**77**] says that $g \in \{1, 2, 3, 4, 6\}$. The isoparametric hypersurfaces with $g = 3$ were classified by Cartan [**19**], cp. also Karcher [**49**], Knarr-Kramer [**52**] and Console-Olmos [**22**]. The 6-dimensional isoparametric hypersurfaces with $g = 6$ were classified by Dorfmeister-Neher [**28**].

It is not difficult to prove that for $g = 1$ one has (up to a scalar factor) an r-sphere $\mathrm{S}^r_\varepsilon \subseteq \mathbb{S}^{r+1}$ of radius ε, and for $g = 2$ one has a Clifford torus $\mathrm{S}^{m_1}_\varepsilon \times \mathrm{S}^{m_2}_\delta \subseteq \mathbb{S}^{m_1+m_2+1}$, for $0 < \varepsilon < 1$ and $\varepsilon^2 + \delta^2 = 1$, cp. Palais-Terng [**82**].

The results of Münzner also show the following. Let m_i denote the multiplicity of the ith principal curvature. After rearranging the indices one has $m_i = m_{i+2}$ for $i = 1, \dots, g$ (indices mod g). The numbers m_1, m_2 are called the *multiplicities* of $\mathcal{F}$. If $g = 1, 3, 6$, then $m_1 = m_2$. If $g = 4$, then either $m_1 = m_2 \in \{1, 2\}$, or $1 \in \{m_1, m_2\}$, or $m_1 + m_2$ is odd, cp. Münzner [**77**], Abresch [**2**].

Stolz [**91**] recently proved the following deep result. Suppose that $g = 4$ and that $2 \leq m_1 < m_2$. Then

$$(m_1, m_2) = (4, 5) \quad \text{or} \quad 2^{\phi(m_1-1)} \text{ divides } m_1 + m_2 + 1,$$

where $\phi(m) = |\{i|\ 1 \leq i \leq m \text{ and } i \equiv 0, 1, 2, 4 \pmod 8\}|$.

8.2. THE GLOBAL GEOMETRY

The fibres of η_{θ_i} are spheres of dimension m_i, for $i = 1, 2$. Conversely, $\mathcal{F}$ can be identified with the normal sphere bundle $\mathrm{S}_{2\sin(\theta_1/2)}(\perp\mathcal{P})$ consisting of all normal vectors of length $2\sin(\theta_1/2)$ (and similarly, $\mathcal{F} \cong \mathrm{S}_{2\sin(\theta_2/2)}(\perp\mathcal{L})$). In fact,

$$\mathcal{F} = \{x \in \mathbb{S}^{r+1}|\ \mathrm{dist}(x, \mathcal{P}) = 2\sin(\theta_1/2)\}$$

and

$$\mathcal{L} = \{x \in \mathbb{S}^{r+1}|\ \mathrm{dist}(x, \mathcal{P}) = 2\sin((\theta_2 - \theta_1)/2)\}.$$

In particular, the isoparametric foliation is completely determined by one focal manifold.

We need the following result.

LEMMA 8.3. *Let $p \in \mathcal{P}$ and let $N \in \perp_p\mathcal{P}$ be a non-zero normal vector. Then the kernel of the Weingarten map A_N is m_2-dimensional. A similar result holds for $\mathcal{L}$; here, the kernel is m_1-dimensional.*

PROOF. This is proved e.g. in Cecil-Ryan [**21**] Ch. 2, Cor. 2.2. □

8.4. THE CORRESPONDING POINT-LINE GEOMETRY

Let $\mathcal{F}$ be an isoparametric hypersurface with focal manifolds $\mathcal{P}$ and $\mathcal{L}$ and $g \geq 2$. Call the elements of $\mathcal{P}$ *points* and the elements of $\mathcal{L}$ lines. A point p and a line ℓ are *incident* if and only if

$$\mathrm{dist}(p, \ell) = 2\sin((\theta_2 - \theta_1)/2).$$

In that case the geodesic arc between p and ℓ meets $\mathcal{F}$ in a unique element $x \in \mathcal{F}$; thus, we can identify $\mathcal{F}$ with the flag set of this geometry. We denote the resulting point-line geometry by $\mathfrak{G}(\mathcal{P}, \mathcal{L})$.

This geometry was introduced in Thorbergsson [**101**]; there, and in Knarr-Kramer [**52**] it is used to classify certain isoparametric submanifolds. The same geometry is also used in Eschenburg-Schröder [**32**] App. If the isoparametric foliation arises from a non-compact Riemannian symmetric space, then this is the same as the building at infinity in the compactification of the symmetric space.

One can show that $\mathfrak{G}(\mathcal{P},\mathcal{L})$ is connected; in fact, $d(x,y) \leq g$ for all elements $x, y \in \mathcal{P} \cup \mathcal{L}$, cp. Eschenburg-Schröder [**32**]. Also, one can show that $\mathfrak{G}(\mathcal{P},\mathcal{L})$ contains no digons, provided that $g \geq 3$ (Thorbergsson, unpublished; see Knarr-Kramer [**52**] for a proof). Let $\ell \in \mathcal{L}$. The point row corresponding to ℓ has a simple description: the sphere of radius $2\sin((\theta_2 - \theta_1)/2)$ around ℓ touches $\mathcal{P}$ along a sphere S, and this set S is precisely the set of all points which are incident with ℓ. In differential-geometric terms, $S \subseteq \mathcal{P}$ is a *curvature sphere.*

For all known examples of isoparametric hypersurfaces (with $g \geq 3$), the geometry $\mathfrak{G}(\mathcal{P},\mathcal{L})$ is a spherical building of rank 2; more precisely, it is a compact generalized g-gon, cp. Thorbergsson [**102**], and, more generally, Immervoll [**48**].

8.5. Isometry groups

Let $\mathrm{Isom}(\mathcal{F})$ denote the group of all isometries of $\mathcal{F}$. Isoparametric hypersurfaces are *rigid*: every isometry of $\mathcal{F}$ is induced by an isometry of $\mathbb{S}^{r+1}$. Thus $\mathrm{Isom}(\mathcal{F}) \subseteq \mathrm{O}(r+2)$. Hsiang-Lawson classified all isoparametric submanifolds where $\mathrm{Isom}(\mathcal{F})$ acts transitively on $\mathcal{F}$. We call such a hypersurface *homogeneous.*

The connected component of $\mathrm{Isom}(\mathcal{F})$ acts also on the focal manifolds $\mathcal{P}$, $\mathcal{L}$; in particular,

$$\mathrm{Isom}(\mathcal{F})^\circ \subseteq \mathrm{Aut}(\mathfrak{G}(\mathcal{P},\mathcal{L})).$$

Conversely, if $G \subseteq \mathrm{O}(r+2)$ is a group which leaves $\mathcal{P}$ or $\mathcal{L}$ invariant, then $G \subseteq \mathrm{Isom}(\mathcal{F})$, cp. 8.2. As the examples by Ferus-Karcher-Münzner show, it is possible that one of the focal manifolds is homogeneous, while $\mathcal{F}$ itself is not homogeneous.

We derive some more general results about transitive actions on focal manifolds.

Lemma 8.6. *Assume that $g \geq 3$. Let $G \subseteq \mathrm{SO}(r+2)$ be a compact connected subgroup. If G acts transitively on the focal manifold $\mathcal{P}$ (or $\mathcal{L}$), then there is no invariant subspace with trivial G-action.*

Proof. Assume otherwise. Let $U \oplus V = \mathbb{R}^{2(m_1+m_2)+2}$ be a decomposition into G-modules such that U is a trivial G-module of positive dimension. Let $(u,v) \in \mathcal{P}$. Then $\mathcal{P} = G \cdot (u,v) = u \times G \cdot v$ is contained in the proper affine subspace $u + V$. But for $g \geq 3$ the focal submanifolds are full, i.e. they are not contained in any proper affine subspace of $\mathbb{R}^{r+2}$. □

Next we note the following. Let $x \in \mathcal{F}$. Then $G \cdot x$ surjects onto $G \cdot \mathrm{pr}_{\mathcal{P}}(x)$. Thus, if $\mathcal{P}$ is G-homogeneous and if $y \in \mathbb{S}^{r+1} \setminus \mathcal{L}$, then the orbit $G \cdot y \subseteq \mathbb{S}^{r+1}$ has at least dimension $\dim(\mathcal{P})$.

Lemma 8.7. *Suppose that $\mathcal{P} = G/G_p \subseteq \mathbb{S}^{r+1}$ is a homogeneous focal manifold. If $x \in \mathbb{S}^{r+1}$ is a point with $\dim(G \cdot x) < \dim(\mathcal{P})$, then $G \cdot x$ is contained in the other focal manifold $\mathcal{L}$.* □

Let $\mathcal{P} = G/G_p$ be a homogeneous focal manifold. Then G_p acts on the normal space $\perp_p\mathcal{P} \subseteq T_p\mathbb{S}^{r+1}$. This is the *normal isotropy representation* of the isotropy group G_p. The normal sphere bundle $S(\perp\mathcal{P})$ can be identified with the isoparametric hypersurface $\mathcal{F}$, and the action of G on this sphere bundle coincides with the action on $\mathcal{F}$. Let $\mathcal{L}$ denote the other focal manifold, and let S denote the curvature sphere corresponding to p. The action of G_p on S coincides with the action of G_p on the normal sphere in $\perp_p\mathcal{P}$. Note that 7.3 applies also to isoparametric hypersurfaces.

LEMMA 8.8. *Let N be a normal almost simple subgroup of G_p, and assume that G_p has no other closed subgroup isomorphic to N. Then*

$$\dim(N) \leq \max\left\{\binom{m_1+1}{2}, \binom{m_2+1}{2}\right\}.$$

PROOF. The proof is the same as in 7.21. □

The problem we will consider is the following. Let $G \subseteq \mathrm{SO}(r+2)$ be a compact connected subgroup which leaves $\mathcal{P}$ or $\mathcal{L}$ invariant, and which acts transitively on $\mathcal{P}$ or on $\mathcal{L}$. We wish to determine all possibilities for $(G, \mathcal{P}, \mathcal{L})$.

The cases $g = 1, 2, 3$ are rather simple. In fact, for $g = 3, 6$, or if $g = 4$ and $m_1 = m_2$, one can show that transitivity of $\mathrm{Isom}(\mathcal{F})$ on one focal manifold implies transitivity on $\mathcal{F}$ itself; the methods developed in Kramer [**54**] for compact polygons carry over to isoparametric hypersurfaces (with some modifications in the smallest case $m_1 = m_2 = 1$), see Wolfrom [**114**].

8.9. THE TOPOLOGY FOR $g = 4$

We consider the remaining case where $g = 4$ and $m_1 \neq m_2$. Then

$$\begin{aligned}\dim(\mathcal{F}) &= 2(m_1 + m_2)\\ \dim(\mathcal{P}) &= 2m_1 + m_2\\ \dim(\mathcal{L}) &= 2m_2 + m_1\end{aligned}$$

If $m_1 \neq m_2$, and $2 \leq m_1, m_2$, then $m_1 + m_2$ is odd. If $m_1 + m_2$ is odd, then

$$\begin{aligned}\mathbf{H}^\bullet(\mathcal{P}) &\cong \mathbf{H}^\bullet(\mathbb{S}^{m_1} \times \mathbb{S}^{m_1+m_2})\\ \mathbf{H}^\bullet(\mathcal{L}) &\cong \mathbf{H}^\bullet(\mathbb{S}^{m_2} \times \mathbb{S}^{m_1+m_2})\\ \mathbf{H}^\bullet(\mathcal{F}) &\cong \mathbf{H}^\bullet(\mathbb{S}^{m_1} \times \mathbb{S}^{m_2} \times \mathbb{S}^{m_1+m_2}).\end{aligned}$$

If $m_1 = 1$ and if m_2 is odd, then

$$\mathbf{H}^\bullet(\mathcal{L}; \mathbb{Q}) \cong \mathbf{H}^\bullet(\mathbb{S}^{2m_2+1}; \mathbb{Q}).$$

Moreover, the following is true. The focal manifold $\mathcal{L}$ is $(m_2 - 1)$-connected, and $\mathcal{P}$ is $(m_1 - 1)$-connected. Let $S \subseteq \mathcal{L}$ be a curvature sphere of dimension m_2, corresponding to $p \in \mathcal{P}$. Then $\pi_{m_2}(\mathcal{L})$ is generated by the inclusion $\mathbb{S}^{m_2} \cong S \subseteq \mathcal{L}$; a similar result holds for $\mathcal{P}$ (note that $\pi_{m_2}(\mathcal{L}) \cong \mathbb{Z}$ if $m_1 < m_2$ and $m_1 + m_2$ is odd and that $\pi_{m_2}(\mathcal{L}) \cong \mathbb{Z}/2$ if $m_1 = 1$ and m_2 is odd).

This follows from Münzner [**77**]. Münzner determined structure constants for the cohomology rings, although he did not write down the resulting rings. See Strauß [**92**] for a thorough discussion.

By the result above, we can apply our classification of homogeneous spaces to focal manifolds.

8.B. The Stiefel manifolds

Let $\mathbb{F} \in \{\mathbb{R}, \mathbb{C}, \mathbb{H}\}$. The Stiefel manifolds $V_2(\mathbb{F}^n)$ can be embedded into spheres as focal manifolds of isoparametric hypersurfaces. Let $G(n)$ be one of the groups $\mathrm{SO}(n)$, $\mathrm{SU}(n)$ or $\mathrm{Sp}(n)$, let V denote the natural $G(n)$-module, and let $\mathbb{F}$ denote the corresponding skew field, $\mathbb{F} = \mathbb{R}, \mathbb{C}, \mathbb{H}$. Consider the action of $G(n)$ on the unit sphere $\mathbb{S}^{2dn-1} \subseteq V \oplus V$, where $d = \dim_{\mathbb{R}} \mathbb{F} = 1, 2, 4$. There are precisely two orbit types. Let $(x, y) \in \mathbb{S}^{2dn-1} \subseteq V \oplus V$. If x, y are $\mathbb{F}$-linearly independent, then $G(n)_{(x,y)} \cong G(n-2)$, and if x, y are $\mathbb{F}$-linearly dependent, then $G_{(x,y)} \cong G(n-1)$.
The union

$$X = \{(x, y) \in \mathbb{S}^{2dn-1} |\ G(n) \cdot x \cong \mathbb{S}^{dn-1}\}$$

of the singular orbits is one focal manifold of the homogeneous isoparametric hypersurface corresponding to this situation, cp. Section 7.E and Section 7.F in the previous chapter. The other focal manifold is the orbit $Y = G(n) \cdot (x, y)$, where $|x|^2 = |y|^2 = 1/2$ and $(x|y) = 0$. However, this orbit is (topologically) not unique; all principal orbits are homeomorphic.

Suppose now that $\mathcal{L} = G(n)/G(n-2) \subseteq \mathbb{R}^{2dn}$ is a homogeneous focal manifold. Assume in addition that the action of $G(n)$ is as above, i.e. that $\mathbb{R}^{2dn} \cong V \oplus V$. The union X of the singular orbits has to be contained in the other focal submanifold $\mathcal{P}$. Now X is a manifold of the same dimension as $\mathcal{P}$, hence $\mathcal{P} = X$. But the isoparametric foliation is uniquely determined by one focal manifold, so $\mathcal{L} = Y$.

LEMMA 8.10. *The homogeneous spaces $G(n)/G(n-2)$ can be realized as homogeneous focal manifolds. If $\mathbb{R}^{2dn} = V \oplus V$, then there is a unique isoparametric foliation in $\mathbb{S}^{2dn-1}$ corresponding to this action.* □

To obtain a complete classification, it remains to show that $\mathbb{R}^{2dn} \cong V \oplus V$. If n is large enough, then there is only one irreducible representation of $G(n)$ of dimension at most $2dn$, the natural one on $V = \mathbb{F}^n$. From 4.12, 4.16, 4.10, 4.14, we see that the precise numbers are as follows: for $\mathbb{F} = \mathbb{R}, \mathbb{C}, \mathbb{H}$ we need $n \geq 10, 6, 5$, respectively. For the low-dimensional cases we use the following fact. According to Lemma 8.6, the $G(n)$-module $\mathbb{R}^{2dn}$ cannot have any trivial factors.

$G(5) = \mathrm{SU}(5)$, $2dn = 20$ By 4.10, the semisimple 20-dimensional $\mathrm{SU}(5)$-$\mathbb{R}$-modules without trivial factors are $V \oplus V$ and ${}^{\mathbb{R}}X_{\lambda_2}$. The orbits in ${}^{\mathbb{R}}X_{\lambda_2}$ yield isoparametric hypersurfaces, but with other orbit types (the multiplicities are $(4, 5)$). Thus in our setting we have $V \oplus V$ as the only possibility.

$G(4) = \mathrm{SU}(4)$, $2dn = 16$ By 4.10, there are no semisimple 16-dimensional $\mathrm{SU}(4)$-$\mathbb{R}$-modules without trivial factors, except for $V \oplus V$.

$G(3) = \mathrm{SU}(3)$, $2dn = 12$ By 4.10, the semisimple 12-dimensional $\mathrm{SU}(3)$-$\mathbb{R}$-modules without trivial factors are $V \oplus V$ and ${}^{\mathbb{R}}X_{2\lambda_1} = \mathrm{S}^2\mathbb{C}^3$. I am indebted to R. Bryant for pointing out that every complex symmetric 3×3-matrix can be diagonalized under the action of $\mathrm{SU}(3)$. Therefore, the isotropy group of every element in $\mathrm{S}^2\mathbb{C}^3$ contains $\mathbb{Z}/2 \oplus \mathbb{Z}/2$, and thus there are no orbits of type $\mathrm{SU}(3)/1$ in $\mathrm{S}^2\mathbb{C}^3$. Thus we have $V \oplus V$ as the only possibility.

$G(9) = \mathrm{SO}(9)$, $2dn = 18$ By 4.12, there are no semisimple 18-dimensional $\mathrm{SO}(9)$-$\mathbb{R}$-modules without trivial factors, except for $V \oplus V$.

$\boxed{G(7) = \mathrm{SO}(7),\ 2dn = 14}$ By 4.12, there are no semisimple 14-dimensional SO(7)-$\mathbb{R}$-modules without trivial factors, except for $V \oplus V$.

$\boxed{G(5) = \mathrm{SO}(5),\ 2dn = 10}$ By 4.12, the semisimple 10-dimensional SO(5)-$\mathbb{R}$-modules without trivial factors are $V \oplus V$ and Ad. In Ad, the principal orbits are $\mathrm{SO}(5)/\mathbb{T}^2$. These orbits belong to a homogeneous isoparametric hypersurface with multiplicities $(2, 2)$, and this excludes this module in our situation. Thus, we we have $V \oplus V$.

$\boxed{G(4) = \mathrm{Sp}(4),\ 2dn = 32}$ By 4.14, there are no semisimple 32-dimensional Sp(4)-$\mathbb{R}$-modules without trivial factors, except for $V \oplus V$.

$\boxed{G(3) = \mathrm{Sp}(3),\ 2dn = 24}$ By 4.14, there are no semisimple 24-dimensional Sp(3)-$\mathbb{R}$-modules without trivial factors, except for $V \oplus V$.

$\boxed{G(2) = \mathrm{Sp}(2),\ 2dn = 16}$ By 4.12, there are no semisimple 16-dimensional Sp(2)-$\mathbb{R}$-modules without trivial factors, except for $V \oplus V$.

$\boxed{G(8) = \mathrm{SO}(8),\ 2dn = 16}$ By 4.12, there are no semisimple 16-dimensional SO(8)-$\mathbb{R}$-modules without trivial factors, except for $V \oplus V$.

$\boxed{G(6) = \mathrm{SO}(6),\ 2dn = 12}$ By 4.12, there are no semisimple 12-dimensional SO(6)-$\mathbb{R}$-modules without trivial factors, except for $V \oplus V$.

Thus we have a complete classification for classical groups acting on Stiefel manifolds. However, there are also the exceptional actions $\mathrm{G}_2/\mathrm{SU}(2) = V_2(\mathbb{R}^7)$, $\mathrm{Spin}(7)/\mathrm{SU}(3) = V_2(\mathbb{R}^8)$, and $\mathrm{Spin}(9)/\mathrm{G}_2 = V_2(\mathbb{O}^2)$. We consider these cases, and some more, in the next section.

8.C. Some sporadic cases

Now we consider some homogeneous spaces of almost simple Lie groups which do not belong to any series. Nevertheless, the ideas are very similar as in the last section.

$\boxed{\mathcal{L} = \mathrm{Spin}(9)/\mathrm{G}_2 = V_2(\mathbb{O}^2)}$ By 4.12, the only 32-dimensional semisimple Spin(9)-$\mathbb{R}$-module without trivial factors is $V \oplus V$, where $V = \mathbb{O} \oplus \mathbb{O}$ is the affine Cayley plane. The orbit types are Spin(9)/Spin(7) (15-dimensional orbits), Spin(9)/Spin(6) (21-dimensional orbits), and $\mathrm{Spin}(9)/\mathrm{G}_2$ (22-dimensional orbits), see Salzmann *et al.* [**85**], Chapter 1. The union X of the singular orbits is the non-homogeneous focal manifold of the Clifford hypersurface with multiplicities $(8, 7)$ (the definite case, cp. Ferus-Karcher-Münzner [**34**] 6.6). Since $X \subseteq \mathcal{P}$ and $\dim(X) = \dim(\mathcal{P})$, we have uniqueness.

$\boxed{\mathcal{P} = \mathrm{Spin}(10)/\mathrm{SU}(5) \text{ or } \mathcal{L} = \mathrm{Spin}(10)/\mathrm{Spin}(7)}$ By 4.16, the only 32-dimensional semisimple Spin(10)-$\mathbb{R}$-module without trivial factors is ${}^{\mathbb{R}}X_{\lambda_4} = \mathbb{C}^{16}$. Inspecting the homogeneous hypersurface with multiplicities $(6, 9)$, we see that there are exactly two singular orbits, of dimensions 24 and 21, respectively; the principal orbits have codimension 1. Thus, the singular orbits are the focal manifolds.

$\boxed{\mathcal{P} = \mathrm{SU}(5)/\mathrm{SU}(2) \times \mathrm{SU}(3) \text{ or } \mathcal{L} = \mathrm{SU}(5)/\mathrm{Sp}(2)}$ By 4.10, the only 20-dimensional semisimple SU(5)-$\mathbb{R}$-modules without trivial factors are $2 \cdot {}^{\mathbb{R}}X_{\lambda_1} = \mathbb{C}^5 \oplus \mathbb{C}^5$ and ${}^{\mathbb{R}}X_{\lambda_2} = \bigwedge^2 \mathbb{C}^5$. In both cases, the orbit structure is known. In our present situation, we have the 20-dimensional simple module. There are exactly 2 singular orbits of type $\mathcal{P} = \mathrm{SU}(5)/\mathrm{SU}(2) \times \mathrm{SU}(3)$ and $\mathcal{L} = \mathrm{SU}(5)/\mathrm{Sp}(2)$.

$\boxed{\mathcal{P} = \mathrm{Spin}(9)/\mathrm{SU}(4)}$ By 4.12, the only 32-dimensional semisimple Spin(9)-$\mathbb{R}$-module without trivial factors is $V \oplus V$, where $V = \mathbb{O} \oplus \mathbb{O}$ is the affine Cayley plane. As observed above, the orbits have dimension $21 = \dim(\mathcal{P})$, 22, or 15. The 15-dimensional orbits have to be contained in the other focal manifold $\mathcal{L}$, and (by counting dimensions) $\mathcal{L}$ is the union of these 15-spheres. Thus we know $\mathcal{L}$, and this determines the isoparametric foliation uniquely.

$\boxed{\mathcal{L} = \mathrm{Spin}(7)/\mathrm{SU}(3) = V_2(\mathbb{R}^8)}$ By 4.12, the only 16-dimensional semisimple Spin(9)-$\mathbb{R}$-module without trivial factors is $\mathbb{R}^8 \oplus \mathbb{R}^8$. The orbits are either 7-spheres (for pairs (cx, sx) with $|x|^2 = 1 = c^2 + s^2$) or Stiefel manifolds $V_2(\mathbb{R}^8)$. As before, the union X of the 7-spheres has to be contained in the other focal manifold $\mathcal{P}$, and thus $X = \mathcal{P}$.

$\boxed{\mathcal{L} = \mathrm{G}_2/\mathrm{SU}(2) = V_2(\mathbb{R}^7)}$ By 4.26, the only 14-dimensional semisimple G_2-$\mathbb{R}$-modules without trivial factors are $\mathbb{R}^7 \oplus \mathbb{R}^7$ and Ad. The orbits in Ad belong to an isoparametric hypersurface $G_2/\mathbb{T}^2$ with 6 distinct principal curvatures and multiplicities $(2, 2)$. This excludes this module. Therefore, we have the standard action of G_2 on pairs of pure octonions. The orbits are either 6-spheres, or Stiefel manifolds $V_2(\mathbb{R}^7)$. As above, the union X of the 6-spheres in $\mathbb{S}^{13}$ is precisely the other focal manifold $\mathcal{P}$.

There are some more pairs (G, H) where G is almost simple. We show that they do not belong to isoparametric foliations.

$\boxed{\mathrm{Sp}(3)/\mathrm{Sp}(1) \times \mathrm{Sp}(1) \text{ and } \mathrm{Sp}(3)/\mathrm{Sp}(1) \times {}^{\mathbb{H}}\rho_{3\lambda_1}(\mathrm{Sp}(1)) \text{ are not possible.}}$ The multiplicities would be $(4, 7)$. However, there is no 24-dimensional semisimple Sp(3)-$\mathbb{R}$-module without trivial factors by 4.14.

$\boxed{\mathcal{P} = \mathrm{E}_6/\mathrm{F}_4 \text{ is not possible.}}$ This follows from Lemma 8.8, because $\dim(\mathrm{F}_4) = 52 > \max\left\{\binom{9}{2}, \binom{8}{2}\right\} = 36$. It follows also from Stolz [**91**]: the multiplicities $(8, 9)$ are not possible.

$\boxed{\mathcal{P} = \mathrm{SU}(6)/\mathrm{Sp}(3) \text{ is not possible.}}$ This follows from Lemma 8.8: $\dim(\mathrm{Sp}(3)) = 21 > \max\left\{\binom{4}{2}, \binom{5}{2}\right\} = 10$.

Note however that SU(6)/Sp(3) and SU(5)/Sp(2) are SU(5)-equivariantly homeomorphic, and that SU(5)/Sp(2) *is* a focal manifold.

8.D. The semisimple case

We begin with the split case. Let $\mathcal{P} = K_1/H_1 \times K_2/H_2$ be a product of homogeneous spheres. We show that in 'most' cases, $\mathcal{P}$ cannot be a focal submanifold of an isoparametric foliation. The idea is exactly the same as in the last chapter. Let $S \subseteq \mathcal{P} = \mathbb{S}^{m_1} \times \mathbb{S}^{m_1+m_2}$ be an m_1-dimensional curvature sphere. Then the composite

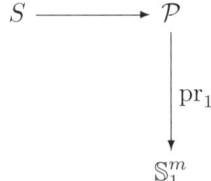

$$\begin{array}{ccc} S & \longrightarrow & \mathcal{P} \\ & & \Big\downarrow \mathrm{pr}_1 \\ & & \mathbb{S}_1^m \end{array}$$

is a homotopy equivalence, hence $S \cap (\{x\} \times \mathbb{S}^{m_1+m_2}) \neq \varnothing$ for all $x \in \mathbb{S}^{m_1}$. The fixed point set $\mathrm{Fix}(H_1, \mathcal{P})$ contains (at least) $\mathbb{S}^0 \times \mathbb{S}^{m_1+m_2} \subseteq \mathcal{P}$. It follows that H_1 fixes all curvature spheres of the type of S globally. But this implies that H_1 fixes $\mathcal{P}$ pointwise, hence $(K_1, H_1) = (\mathrm{Sp}(1), 1)$.

LEMMA 8.11. *If $\mathcal{P} = K_1/H_1 \times K_2/H_2$ is a product of homogeneous spheres, then one of the factors is $\mathbb{S}^3$ with the regular* $\mathrm{Sp}(1)$*-action.* □

To exclude the homogeneous spheres $K_2/H_2 = \mathrm{SO}(m_2+4)/\mathrm{SO}(m_2+3)$, for $m_2 \geq 2$, we apply Lemma 8.8 to H_2. In fact $\dim(H_2) \leq \max\left\{\binom{4}{2}, \binom{m_2+1}{2}\right\}$. Note that $\mathbb{S}^3 \times \mathbb{S}^4$ cannot be a focal manifold, since $\pi_3(\mathbb{S}^3 \times \mathbb{S}^4) \neq \mathbb{Z}/2$, so $m_2 = 1$ is also excluded.

Suppose that $\mathcal{P} = (\mathrm{SU}(r+1)/\mathrm{SU}(r)) \times \mathbb{S}^3$. Then $m_2 = 2r-2$. Here we cannot use 8.8. However, $\mathrm{SU}(r)$ cannot act non-trivially on $\mathbb{R}^{2r-1}$, provided that $r \neq 2, 4$, cp. 4.10. Thus, $\mathrm{SU}(r)$ cannot act non-trivially on $\perp_p \mathcal{P}$ for $n \neq 2, 4$, and this excludes these groups. Similarly, the cases $\mathcal{P} = (\mathrm{G}_2/\mathrm{SU}(3)) \times \mathbb{S}^3$ or $\mathcal{P} = (\mathrm{Spin}(7)/\mathrm{G}_2) \times \mathbb{S}^3$ are not possible. We are left with the cases

$$
\begin{gathered}
(\mathrm{Sp}(n)/\mathrm{Sp}(n-1)) \times \mathbb{S}^3, \ n \geq 2 \\
(\mathrm{Spin}(9)/\mathrm{Spin}(7)) \times \mathbb{S}^3 \\
(\mathrm{SU}(5)/\mathrm{SU}(4)) \times \mathbb{S}^3 \\
(\mathrm{SU}(3)/\mathrm{SU}(2)) \times \mathbb{S}^3
\end{gathered}
$$

In the non-split case we have also to consider the pairs $(\mathrm{Sp}(n) \times \mathrm{Sp}(2), \mathrm{Sp}(n-1) \cdot \mathrm{Sp}(1))$ and $(\mathrm{Sp}(n) \times \mathrm{SU}(3), \mathrm{Sp}(n-1) \cdot \mathrm{Sp}(1))$. By the same reasoning as above, one sees that these groups cannot occur for $n \geq 4$, since then $\mathrm{Sp}(n-1)$ cannot act non-trivially on real vector spaces of dimension less than $4(n-1)$. We have to consider the pairs

$$
\begin{gathered}
(\mathrm{Sp}(3) \times \mathrm{Sp}(2), \mathrm{Sp}(2) \cdot \mathrm{Sp}(1)) \\
(\mathrm{Sp}(3) \times \mathrm{SU}(3), \mathrm{Sp}(2) \cdot \mathrm{Sp}(1)) \\
(\mathrm{Sp}(2) \times \mathrm{SU}(3), \mathrm{Sp}(1) \cdot \mathrm{Sp}(1)),
\end{gathered}
$$

and some more cases.

$\boxed{\mathcal{P} = \mathrm{Spin}(9)/\mathrm{Spin}(7) \times \mathbb{S}^3 \text{ is not possible.}}$ Otherwise we have a 32-dimensional $\mathrm{Spin}(9)$-$\mathbb{R}$-module. The low-dimensional irreducible representations have dimension 9 and 15, so $\mathbb{R}^{32} = V \oplus W$ splits off the $\mathrm{Spin}(9)$-module $V = \mathbb{O} \oplus \mathbb{O}$.

Suppose that W splits off the natural $\mathrm{SO}(9)$-$\mathbb{R}$-module $X = \mathbb{R}^9$. The remaining 7-dimensional $\mathrm{Spin}(9)$-module has to be trivial, cp. 4.12. Moreover, $\mathrm{Cen}_{\mathrm{SO}(32)}(\mathrm{Spin}(9)) \cong \mathrm{SO}(7)$. Therefore we see that $\mathcal{P} = \mathbb{S}^{15} \times \mathbb{S}^3$ factors in such a way that $\mathbb{S}^{15} \subseteq V \oplus W$ and $\mathbb{S}^3 \subseteq \mathbb{R}^7$. Let $p = (v, w, x) \in \mathcal{P} \subseteq \mathcal{V} \oplus W \oplus \mathbb{R}^7$ and let $N \in \perp_p \mathcal{P} \cap \mathbb{R}^7$ be a non-zero normal vector (i.e. choose a non-zero vector in $\mathbb{R}^7$ perpendicular to the $\mathbb{S}^3$-orbit of p). Then $N_{(v',w',x)} = N$ defines a normal vector field for all $(v', w') \in \mathbb{S}^{15}$. The Weingarten map is $A_N X = 0$ for all $X \in T_{(v,w)}\mathbb{S}^{15} \subseteq T(v, w, x)\mathcal{P}$. Thus, the kernel of A_N is at least 15-dimensional, contradicting 8.3.

If W is a trivial $\mathrm{Spin}(9)$-module, then by a similar argument A_N has a kernel of dimension at least 15, provided that $N \subseteq W$ is a normal vector.

Finally, suppose that $W \cong V$. Then $\mathrm{Cen}_{\mathrm{SO}(32)}(\mathrm{Spin}(9))^\circ \cong \mathrm{SO}(2)$; thus, there is no room left for $\mathrm{Sp}(1)$.

$\boxed{\mathcal{P} = \mathrm{SU}(5)/\mathrm{SU}(4) \times \mathbb{S}^3 \text{ is not possible.}}$ The corresponding isoparametric hypersurface would have multiplicities $(3, 6)$, contradicting Stolz' result.

$\boxed{\mathcal{P} = \mathrm{SU}(3)/\mathrm{SU}(2) \times \mathbb{S}^3 \text{ is not possible.}}$ Suppose otherwise. We have a 12-dimensional SU(3)-$\mathbb{R}$-module. From 4.10 we see that $\mathbb{R}^{12} \cong V \oplus W$, where $V = \mathbb{C}^3$ is the natural module.

If $\mathbb{R}^{12} \cong V \oplus V$, then $\mathrm{Cen}_{\mathrm{SO}(12)}(\mathrm{SU}(3)) \cong \mathrm{SU}(2)$; thus, we know the action of $\mathrm{SU}(3) \times \mathrm{SU}(2)$ on $\mathbb{R}^{12}$. However, the principal orbits of this action are isoparametric hypersurfaces with multiplicities $(2, 1)$, and this excludes the module $V \oplus V$.

Otherwise, W splits off a non-zero trivial SU(3)-module. By the same argument as above, we see that then the Weingarten map of $\mathcal{P}$ has a 5-dimensional kernel. However, the multiplicities in this case are $(3, 2)$, and thus we have again a contradiction to 8.3.

$\boxed{\mathcal{P} = \mathrm{Sp}(3) \times \mathrm{Sp}(2)/\mathrm{Sp}(2) \cdot \mathrm{Sp}(1) \text{ is not possible.}}$ Otherwise we have a 24-dimensional Sp(3)-$\mathbb{R}$-module. Then $\mathbb{R}^{24}$ splits off the natural Sp(3)-module V by 4.14.

If $\mathbb{R}^{24} \cong V \oplus V$ (as a Sp(3)-module), then $\mathrm{Cen}_{\mathrm{SO}(24)}(\mathrm{Sp}(3))^\circ \cong \mathrm{Sp}(2)$, thus we know the orbit structure. The principal orbits are isoparametric hypersurfaces with multiplicities $(4, 7)$, but both focal manifolds have other orbit types than the one which we consider here.

Otherwise, $\mathbb{R}^{24}$ splits off a 12-dimensional trivial module, cp. 4.14, and therefore $\mathrm{Cen}_{\mathrm{SO}(24)}(\mathrm{Sp}(3))^\circ \cong \mathrm{Sp}(1) \times \mathrm{SO}(12)$. However, $\mathbb{R}^{12}$ cannot be a semisimple Sp(2)-module without trivial factors, cp. 4.12. Thus, $\mathbb{R}^{24}$ cannot be a semisimple $\mathrm{Sp}(3) \times \mathrm{Sp}(2)$-module without trivial factors.

$\boxed{\mathcal{P} = \mathrm{Sp}(3) \times \mathrm{SU}(3)/\mathrm{Sp}(2) \cdot \mathrm{Sp}(1) \text{ is not possible.}}$ Here, the multiplicities would be $(5, 6)$, contradicting Stolz [**91**].

$\boxed{\mathcal{P} = \mathrm{Sp}(2) \times \mathrm{SU}(3)/\mathrm{Sp}(1) \cdot \mathrm{Sp}(1) \text{ is not possible.}}$ Here, the multiplicities would be $(5, 2)$. We decompose $\mathbb{R}^{16}$ as a semisimple Sp(2)-$\mathbb{R}$-module. From 4.12 we see that $\mathbb{R}^{16}$ splits off the natural Sp(2)-module V. If $\mathbb{R}^{16} \cong V \oplus V$, then $\mathrm{Cen}_{\mathrm{SO}(16)}(\mathrm{Sp}(2)) \cong \mathrm{Sp}(2)$; however, there is no embedding $\mathrm{SU}(3) \subseteq \mathrm{Sp}(2)$, cp. 4.10. Put $\mathbb{R}^{16} = V \oplus W$. Then $\mathrm{SU}(3) \subseteq \mathrm{Cen}_{\mathrm{SO}(16)}(\mathrm{Sp}(2))^\circ \subseteq \mathrm{Sp}(1) \times \mathrm{SO}(8)$. Thus, V is a trivial SU(3)-module. Now W is an 8-dimensional non-trivial SU(3)-$\mathbb{R}$-module. By 4.10, W splits off the natural SU(3)-module. We have a remaining 2-dimensional $\mathrm{Sp}(2) \times \mathrm{SU}(3)$-$\mathbb{R}$-module, which has to be trivial. Thus, $\mathbb{R}^{16}$ cannot be a semisimple $\mathrm{Sp}(2) \times \mathrm{SU}(3)$-module without trivial factors.

In the split case, only the infinite series $\mathbb{S}^{4n-1} \times \mathbb{S}^3 = \mathrm{Sp}(n)/\mathrm{Sp}(n-1) \times \mathbb{S}^3$ with multiplicities $(3, 4n-4)$ remains; in the non-split case, the series $\mathrm{Sp}(n) \times \mathrm{Sp}(2)/\mathrm{Sp}(n-1) \cdot \mathrm{Sp}(1) \cdot \mathrm{Sp}(1)$ with multiplicities $(4, 4n-5)$ remains.

8.E. The $(4, 4n-5)$- and the $(3, 4n-4)$-series

Now we consider the remaining two infinite series. We start with the series

$$\mathcal{P} = \mathrm{Sp}(n) \times \mathrm{Sp}(2)/\mathrm{Sp}(n-1) \cdot \mathrm{Sp}(1) \cdot \mathrm{Sp}(1)),$$

cp. Section 7.F. Here, the multiplicities are $(4, 4n-5)$ and $G = \mathrm{Sp}(n) \times \mathrm{Sp}(2)$. Consider the action of $\mathrm{Sp}(n-1)$ on the normal space $\perp_p \mathcal{P} \cong \mathbb{R}^{4n-4}$.

If this action is trivial, then $\mathrm{Sp}(n-1)$ fixes a curvature sphere in the other focal manifold $\mathcal{L}$. The normal space of $\mathcal{L}$ is 5-dimensional. If $n-1 \geq 3$, then $\mathrm{Sp}(n-1)$ acts also trivially on the normal space $\perp_\ell \mathcal{L}$, since there is no 5-dimensional non-trivial $\mathrm{Sp}(n-1)$-$\mathbb{R}$-module, cp. 4.14. It follows then that $\mathrm{Sp}(n-1)$ fixes every point in $\mathcal{P}$, which is absurd.

Thus, $\mathrm{Sp}(n-1)$ acts non-trivially on $\perp_p \mathcal{P} \cong \mathbb{R}^{4n-4}$. If $n-1 \geq 3$, then this is the natural module by 4.14, and thus the action of G on the normal sphere bundle $\mathrm{S}(\perp \mathcal{P}) \cong \mathcal{F}$ is transitive. It follows that the action of G on $\mathcal{L}$ is also transitive. In Section 8.B we classified all homogeneous focal manifolds $\mathcal{L}$ belonging to isoparametric foliations with multiplicities $(4, 4n-5)$, for $n \geq 3$; then $\mathcal{L}$ is a Stiefel manifold $V_2(\mathbb{H}^n)$. Note that the action of G on $\mathcal{L}$ is not irreducible; the normal subgroup $\mathrm{Sp}(n)$ acts already transitively on $\mathcal{L}$. We have established the uniqueness of the isoparametric foliation for $n \geq 4$. We consider the cases $n = 2, 3$ separately.

$\boxed{n=3}$ Then we have a 24-dimensional $\mathrm{Sp}(3)$-module. Thus $\mathbb{R}^{24} = V \oplus W$, and $V = \mathbb{H}^3$ is the natural $\mathrm{Sp}(3)$-module, cp. 4.14.

If W is a trivial $\mathrm{Sp}(3)$-$\mathbb{R}$-module, then $\mathrm{Cen}_{\mathrm{SO}(24)}(\mathrm{Sp}(3)) \cong \mathrm{Sp}(1) \times \mathrm{SO}(12)$. Thus $\mathrm{Sp}(2) \subseteq \mathrm{SO}(12)$, and $W \cong \mathbb{R}^{12}$ has to be a semisimple $\mathrm{Sp}(2)$-module without trivial factors. By 4.12, this is not possible.

Thus $\mathbb{R}^{24} = V \oplus V$. Then $\mathrm{Cen}_{\mathrm{SO}(24)}(\mathrm{Sp}(3)) \cong \mathrm{Sp}(2)$. Thus, we have uniqueness of the G-action. The principal G-orbits are isoparametric.

$\boxed{n=2}$ Then we have a 16-dimensional $\mathrm{Sp}(2)$-$\mathbb{R}$-module. Thus $\mathbb{R}^{16} = V \oplus W$, and $V = \mathbb{H}^2$ is the natural $\mathrm{Sp}(2)$-module, cp. 4.12.

If W is a trivial $\mathrm{Sp}(2)$-$\mathbb{R}$-module, then $\mathrm{Cen}_{\mathrm{SO}(16)}(\mathrm{Sp}(2)) \cong \mathrm{Sp}(1) \times \mathrm{SO}(8)$. Then the other factor $\mathrm{Sp}(2)$ is contained in $\mathrm{SO}(8)$; as in the case $n=3$ above, we conclude that it acts as $\mathrm{Sp}(2)$ on $\mathbb{H}^2 \cong \mathbb{R}^8$. Then the G-orbits are either products of spheres $\mathbb{S}^7 \times \mathbb{S}^7$ or spheres $\mathbb{S}^7$; thus, $\mathcal{P}$ cannot be an orbit in this module.

Suppose that $W = \mathbb{R}^5 \oplus \mathbb{R}^3$ decomposes into the natural $\mathrm{SO}(5)$-$\mathbb{R}$-module and a trivial 3-dimensional module. Then $\mathrm{Cen}_{\mathrm{SO}(16)}(\mathrm{Sp}(2)) \cong \mathrm{Sp}(1) \times \mathrm{SO}(3)$, and there is no more room for the other factor $\mathrm{Sp}(2)$.

Finally, suppose that $\mathbb{R}^{16} = V \oplus V$. Then $\mathrm{Cen}_{\mathrm{SO}(16)}(\mathrm{Sp}(2)) \cong \mathrm{Sp}(2)$, and we have uniqueness of the action of G. The principal orbits are isoparametric.

PROPOSITION 8.12. *Each of the homogeneous spaces in the $(4, 4n-5)$-series, $n \geq 2$, is in a unique way a focal manifold of an isoparametric foliation.* □

Now we consider the series

$$\mathcal{P} = \mathrm{Sp}(n)/\mathrm{Sp}(n-1) \times \mathrm{Sp}(1).$$

The multiplicities are $(3, 4n-4)$. First we describe the known example.

8.13. EXAMPLE

Let

$$\mathcal{P} = \{(x, xa) \in \mathbb{H}^n \oplus \mathbb{H}^n |\ x \in \mathbb{H}^n,\ a \in \mathbb{H},\ |x|^2 = 1/2,\ |a|^2 = 1\}$$
$$\mathcal{L} = \{(u, v) \in \mathbb{H}^n \oplus \mathbb{H}^n |\ (u|v) = 0,\ |u|^2 + |v|^2 = 1\}$$

(here $(-|-)$ denotes the standard positive definite hermitian form on $\mathbb{H}^n$). These are focal manifolds of an isoparametric hypersurface with $g = 4$ distinct principal curvatures. The multiplicities are $(3, 4n-4)$. The incidence in the corresponding point-line geometry is as follows. A point $p = (x, xa)$ is incident with the line $\ell = (u, v)$ if and only if

$$x = \frac{1}{\sqrt{2}}(u + v\bar{a}).$$

The group $\mathrm{Sp}(n) \times \mathrm{Sp}(1) \times \mathrm{Sp}(1)$ acts as a group of automorphisms and isometries on the isoparametric foliation (the group $\mathrm{Sp}(n)$ as a matrix group from the left, the group $\mathrm{Sp}(1) \times \mathrm{Sp}(1)$ as multiplication by pairs of scalars from the right), and the action is transitive on the point space $\mathcal{P}$. The subgroup $\mathrm{Sp}(n) \times \mathrm{Sp}(1)$ (where $\mathrm{Sp}(1)$ is any of the two factors) acts still transitively on the points.

Now we get back to the unknown homogeneous focal manifold. If the action of $\mathrm{Sp}(n-1)$ on $\perp_p\mathcal{P} \cong \mathbb{R}^{4n-3}$ is trivial, then $\mathrm{Sp}(n-1)$ has to act on the space $\perp_\ell\mathcal{L} \cong \mathbb{R}^4$, which is impossible for $n \geq 3$. Thus we have a non-trivial action of $\mathrm{Sp}(n-1)$ on $\perp_p\mathcal{P} \cong \mathbb{R}^{4n-3}$, provided that $n \geq 3$. If $n \geq 4$, then the only possibility is that $\mathbb{R}^{4n-4} \cong V \oplus \mathbb{R}$ is the natural $\mathrm{Sp}(n-1)$-module V plus a 1-dimensional trivial module. Thus, we see that the normal bundle of $\mathcal{P}$ is

$$\perp\mathcal{P} = (\mathrm{Sp}(n) \times \mathrm{Sp}(1)) \times_{\mathrm{Sp}(n-1)} (V \oplus \mathbb{R}).$$

The isoparametric hypersurface is the unit sphere bundle of the normal bundle $\perp\mathcal{P}$.

Consider the $\mathrm{Sp}(n)$-$\mathbb{R}$-module $\mathbb{R}^{8n}$. From 4.14 we see that $\mathbb{R}^{8n} \cong V \oplus W$ splits off the natural module $V = \mathbb{H}^n$ (this holds for all $n \geq 2$).

Suppose that W is the trivial module. Let $(v, w) \in \mathcal{P} \subseteq V \oplus W$. Then $w \neq 0$ and $\mathrm{Sp}(n) \cdot (v, w)$ is a sphere $\mathbb{S}^{4n-1}$. The normal isotropy representation of $\mathrm{Sp}(n-1)$ on $\perp_p(\mathrm{Sp}(n) \cdot (v, w)) \cong \mathbb{R} \oplus W$ is trivial; it follows that the normal isotropy representation of $\mathrm{Sp}(n-1)$ on $\perp_p\mathcal{P}$ is trivial as well. But this is impossible for $n \geq 3$.

If $n \geq 3$, then W has to be trivial or isomorphic to the natural module V, cp. 4.14. Thus, $\mathbb{R}^{8n} \cong V \oplus V$ as a $\mathrm{Sp}(n)$-module, provided that $n \geq 3$.

$\boxed{n = 2}$ Then it is also possible that $W = X \oplus \mathbb{R}^3$ splits off the 5-dimensional $\mathrm{SO}(5)$-$\mathbb{R}$-module X. Suppose that we have this module. Then $\mathrm{Cen}_{\mathrm{SO}(16)}(\mathrm{Sp}(2)) \cong \mathrm{Sp}(1) \times \mathrm{SO}(3)$. Let $p = (w, x, z) \in \mathcal{P} \subseteq \mathcal{V} \oplus W \oplus \mathbb{R}^3$. If $w \neq 0 \neq x$, then the $\mathrm{Sp}(2)$-stabilizer of p is trivial. But for $p \in \mathcal{P}$, the stabilizer is $\mathrm{Sp}(1)$. Thus $x = 0$, and $\mathcal{P} \subseteq V \oplus \mathbb{R}^3$, which is impossible. Thus $\mathbb{R}^{16} = V \oplus V$ as a $\mathrm{Sp}(2)$-module.

We have established for all $n \geq 2$ that $\mathbb{R}^{8n} = V \oplus V$ as a $\mathrm{Sp}(n)$-module, and

$$\mathbb{S}^3 \subseteq \mathrm{Cen}_{\mathrm{SO}(8n)}(\mathrm{Sp}(n)) \cong \mathrm{Sp}(2).$$

Let $p = (x, y) \in \mathcal{P} \subseteq V \oplus V$. The $\mathrm{Sp}(n)$-orbit of p is a sphere, therefore (x, y) have to be $\mathbb{H}$-linearly dependent. Put

$$\widehat{\mathcal{P}} = \{(xc, xs) |\ x \in \mathbb{S}^{4n-1}, c, s \in \mathbb{H}, |c|^2 + |s|^2 = 1\}.$$

Then $\mathcal{P} \subseteq \widehat{\mathcal{P}}$. As in Section 7.F, consider the map

$$\phi : \widehat{\mathcal{P}} \longrightarrow \mathbb{S}^4, \quad (xc, xs) \longmapsto (|xc|^2 - |xs|^2, 2\bar{c}s).$$

The fibres of ϕ are precisely the $\mathrm{Sp}(n)$-orbits in $\widehat{\mathcal{P}}$. The group $\mathrm{Cen}_{\mathrm{SO}(8n)}(\mathrm{Sp}(n)) \cong \mathrm{Sp}(2)$ permutes these fibres and acts as $\mathrm{SO}(5)$ on the image $\mathbb{S}^4$ of ϕ; in other words, ϕ is $\mathrm{Sp}(2)$-equivariant. Now $\mathbb{S}^3 \subseteq \mathrm{Sp}(2)$ has an orbit on $\mathcal{P}$ which meets every $\mathrm{Sp}(n)$-orbit precisely once, and which is homeomorphic to $\mathbb{S}^3$. Thus we have to consider subgroups isomorphic to $\mathbb{S}^3$ in $\mathrm{SO}(5)$. By 4.10, there is just one conjugacy class of such groups, $SU(2) \subseteq SO(4) \subseteq SO(5)$. We lift this group into $Sp(2)$ to obtain $\mathbb{S}^3 \subseteq Sp(2)$.

So far, we have determined the subgroup $\mathbb{S}^3 \subseteq Sp(2)$; it is conjugate to $Sp(1) \subseteq Sp(2)$, and the action of $G = Sp(n) \times Sp(1)$ on $\mathbb{R}^{8n}$. The problem is that there are many G-orbits in $\mathbb{S}^{8n-1}$ which are homeomorphic to $\mathcal{P}$, and we have to find the right one. Note that for each orbit homeomorphic to $\mathcal{P}$, the normal bundle has the form described above. Therefore we know the orbits which are contained in the hypersurfaces, or in $\mathcal{P}$; they are either homeomorphic to $\mathcal{P}$ (and thus $(4n+2)$-dimensional), or $(8n-3)$-dimensional (for pairs (x, y), where x, y are linearly independent, but $(x|y) \neq 0$). However, there are also $(8n-6)$-dimensional orbits (for pairs (u, v), where u, v are linearly independent, with $(u|v) = 0$) and $(4n-1)$-dimensional orbits (for $x = 0$ or $y = 0$). These orbits have to be contained in the other focal manifold $\mathcal{L}$. In fact, $\mathcal{L}$ consists entirely of these orbits, because the union of these orbits is a manifold of the same dimension as $\mathcal{L}$ — here we look at our model as in 8.13. Thus, $\mathcal{L}$ is unique, and this establishes the uniqueness of the isoparametric foliation.

PROPOSITION 8.14. *Each of the spaces in the $(3, 4n-4)$-series, $n \geq 2$, is in a unique way a focal manifold of an isoparametric foliation.* □

8.F. Summary

In this chapter, we have established the following result.

THEOREM 8.15. *Let $\mathcal{F} \subseteq \mathbb{S}^{r+1}$ be an isoparametric hypersurface with four distinct principal curvatures and multiplicities (m_1, m_2). Let $\mathcal{P}$, $\mathcal{L}$ denote the focal manifolds of $\mathcal{F}$. Suppose that $G \subseteq SO(r+2)$ is a compact connected subgroup which acts transitively on the focal manifold $\mathcal{P}$ and assume that $m_1 \geq 3$.*

There are the following possibilities for the isoparametric foliation (and no others).

- $\mathcal{F}$ ***is homogeneous.*** *Then $\mathcal{F}$ belongs to the homogeneous series with multiplicities $(n-2, 1)$, for $n \geq 5$, or $(2n-3, 2)$, for $n \geq 3$, or $(4n-5, 4)$, or dually $(4, 4n-5)$, for $n \geq 2$, or has multiplicities $(4, 5)$, $(5, 4)$, $(6, 9)$, or $(9, 6)$.*
- $\mathcal{F}$ ***is not homogeneous.*** *Then $\mathcal{F}$ is of Clifford type. Either it belongs to the non-homogeneous Clifford series with multiplicities $(3, 4n-4)$, for $n \geq 2$, or $(m_1, m_2) = (7, 8)$, and $(\mathcal{P}, \mathcal{L}, \mathcal{F}) = (M_+, M_-, M)$ is dual to the definite isoparametric foliation of Clifford type with multiplicities $(8, 7)$ (cp. Ferus-Karcher-Münzner* [**34**] *for the terminology).*

COROLLARY 8.16. *Let $\mathcal{F}$ be an isoparametric hypersurface with four distinct principal curvatures. If both focal manifolds $\mathcal{P}$ and $\mathcal{L}$ are homogeneous, then $\mathcal{F}$ is homogeneous.* □

The following result was recently proved by Wolfrom.

THEOREM 8.17 (Wolfrom). *Let $\mathcal{F}$ be an isoparametric hypersurface with $g = 3, 4, 6$, and assume that $m_1 = 2$. If the focal manifold $\mathcal{P}$ is homogeneous, then $\mathcal{F}$ is homogeneous* [**114**].

COROLLARY 8.18. *Let $\mathcal{F}$ be an isoparametric hypersurface, and assume that one focal manifold is homogeneous. Then either $\mathcal{F}$ is homogeneous, or $\mathcal{F}$ is of Clifford type, with $g = 4$ and $(m_1, m_2) = (8, 7)$ or $(m_1, m_2) = (3, 4k)$.*

PROOF. We may assume that $\mathcal{P}$ is homogeneous, and that $g = 4, 6$. If $m_1 \geq 3$, then the result follows from our Theorem 8.15 above, and for $m_1 = 2$ it follows from Wolfrom's Theorem 8.17. If $m_1 = 1$, then the results of Takagi [**97**] and Dorfmeister-Neher [**28**] apply. □

Finally, we state Immervoll's theorem.

THEOREM 8.19 (Immervoll). *The point-line geometry associated to an isoparametric hypersurface with $g = 4$ is a compact connected and smooth generalized quadrangle* [**48**].

The following difficult problem is open.

8.20. CONJECTURE *Every isoparametric hypersurface is either homogeneous, or of Clifford type.*

Looking back at our proof of Theorem 8.15, we note that we have obtained the following classification of compact connected transitive groups. Let $\mathcal{F}$ be an isoparametric hypersurface as above, and let $G \subseteq \mathrm{SO}(r+2)$ be a compact connected group which acts transitively on $\mathcal{P}$. There are precisely the following possibilities for G.

8.21. MULTIPLICITIES $(n-2, 1)$, FOR $n \geq 5$

Then $\mathcal{P} = V_2(\mathbb{R}^n)$ is a Stiefel manifold, and $G = \mathrm{SO}(n)$ or $G = \mathrm{SO}(n) \cdot \mathrm{SO}(2)$, where $\mathrm{SO}(n) \cap \mathrm{SO}(2) = \mathrm{SO}(n) \cap \{\pm 1\}$.

For $n = 8$, there are the additional possibilities $G = \mathrm{Spin}(7)$ and $G = \mathrm{Spin}(7) \cdot \mathrm{SO}(2)$, where $\mathrm{Spin}(7) \cap \mathrm{SO}(2) = \{\pm 1\}$, and for $n = 7$ there is the additional possibility $G = \mathrm{G}_2$ or $G = \mathrm{G}_2 \cdot \mathrm{SO}(2)$, where $\mathrm{G}_2 \cap \mathrm{SO}(2) = 1$.

The action of G is transitive on $\mathcal{L}$ and $\mathcal{F}$ if and only if G contains the second factor $\mathrm{SO}(2)$.

8.22. MULTIPLICITIES $(2n-3, 2)$, FOR $n \geq 3$

Then $\mathcal{P} = V_2(\mathbb{C}^n)$ is a Stiefel manifold, and G is one of the groups $\mathrm{SU}(n)$, $\mathrm{U}(n)$, $\mathrm{SU}(n) \cdot \mathrm{SU}(2)$, $\mathrm{U}(n) \cdot \mathrm{U}(2)$, where $\mathrm{SU}(n) \cap \mathrm{SU}(2) = \mathrm{SU}(n) \cap \{\pm 1\}$ and $\mathrm{U}(n) \cap \mathrm{SU}(2) = \{\pm 1\}$.

The group G acts transitively on $\mathcal{L}$ or $\mathcal{F}$ only if G contains the $\mathrm{SU}(2)$ factor.

8.23. MULTIPLICITIES $(4, 4n-5)$ OR $(4n-5, 4)$, FOR $n \geq 2$

Then $\mathcal{L} = V_2(\mathbb{H}^n)$ is a Stiefel manifold, and G is of the form $G = \mathrm{Sp}(n) \cdot K$, where K is a connected subgroup of $\mathrm{Sp}(2)$, and $\mathrm{Sp}(2) \cap \mathrm{Sp}(n) = \{\pm 1\}$ (it is not difficult to determine all connected subgroups of $\mathrm{Sp}(2)$). The action of G is transitive on the focal manifold $\mathcal{L}$, and transitive on $\mathcal{P}$ or $\mathcal{F}$ if and only if $K = \mathrm{Sp}(2)$.

8.24. MULTIPLICITIES $(4, 5)$ OR $(5, 4)$

The group is either $G = \mathrm{SU}(5)$ or $G = \mathrm{U}(5)$. In each case, G acts transitively on $\mathcal{P}$, $\mathcal{L}$ and $\mathcal{F}$.

8.25. MULTIPLICITIES $(9, 6)$ OR $(6, 9)$

If G acts transitively on $\mathcal{L}$, then $G = \mathrm{Spin}(10)$ or $G = \mathrm{Spin}(10) \cdot \mathrm{U}(1)$, and G acts transitively on $\mathcal{L}$ and $\mathcal{F}$. The groups $G = \mathrm{Spin}(9)$ and $G = \mathrm{Spin}(9) \cdot \mathrm{U}(1)$ act transitively on $\mathcal{P}$, but not on $\mathcal{L}$ or $\mathcal{F}$.

8.26. Multiplicities $(3, 4n-4)$, for $n \geq 2$

Then $\mathcal{P} = \mathbb{S}^3 \times \mathbb{S}^{4n-5}$, and G is of the form $G = \mathrm{Sp}(n) \cdot K$, where K is a connected subgroup of $\mathrm{Sp}(1) \times \mathrm{Sp}(1)$, containing a subgroup isomorphic to $\mathrm{Sp}(1)$. The group G acts transitively on $\mathcal{P}$, but not on $\mathcal{L}$ or $\mathcal{F}$.

8.27. Multiplicities $(7, 8)$

If G acts transitively on $\mathcal{P}$, then $G = \mathrm{Spin}(9)$ or $G = \mathrm{Spin}(9) \cdot \mathrm{SO}(2)$, and G does not act transitively on $\mathcal{L}$ or $\mathcal{F}$. Here, $\mathcal{P} = V_2(\mathbb{O}^2) = M_+$.

Bibliography

[1] P. Abramenko, *Twin buildings and applications to S-arithmetic groups.* LNM 1641, Springer-Verlag, Berlin (1996) x+123 pp. MR 99k:20060

[2] U. Abresch, *Isoparametric hypersurfaces with four or six distinct principal curvatures.* Math. Ann. 264 (1983) 283–302. MR 85g:53052b

[3] W. Ballmann, M. Gromov, and V. Schröder, *Manifolds of nonpositive curvature.* Birkhäuser, Boston, MA. (1985) vi+263 pp. MR 87h:53050

[4] H. Biller, *Actions of compact groups on spheres and on generalized quadrangles.* Ph.D. Thesis, Stuttgart: Math. Fak., Univ. Stuttgart, (1999) xxvi+195 pp.

[5] O. Bletz, *Ein Beweis für die Lokalkompaktheit der Automorphismengruppen kompakter Polygone. Diplomarbeit,* Würzburg: Math. Fak., Univ. Würzburg, (1999) 40 pp.

[6] R. Bödi and M. Joswig, *Tables for an effective enumeration of real representations of quasi-simple Lie groups.* Sem. Sophus Lie 3 (1993) 239–253. MR 95f:22003

[7] R. Bödi and L. Kramer, *On homomorphisms between generalized polygons.* Geom. Dedicata 58 (1995) 1–14. MR 96k:51017

[8] A. Borel, *Some remarks about Lie groups transitive on spheres and tori.* Bull. Amer. Math. Soc. 55 (1949) 580–587. MR 10,680c

[9] A. Borel, *Sur la cohomologie des espaces fibrés principaux et des espaces homogènes de groupes de Lie compacts.* Ann. of Math. 57 (1953) 115–207. MR 14,490e

[10] A. Borel, *Sur l'homologie et la cohomologie des groupes de Lie compacts connexes.* Amer. J. Math. 76 (1954) 273–342. MR 16,219b

[11] A. Borel, *Seminar on transformation groups.* With contributions by G. Bredon, E. E. Floyd, D. Montgomery, R. Palais, Princeton Univ. Press, Princeton, NJ. (1960) vii+245 pp. MR 22#7129

[12] A. Borel, *Topics in the homology theory of fibre bundles.* LNM 36, Springer-Verlag, Berlin-New York (1967) 95 pp. MR 36#4559

[13] A. Borel and J. De Siebenthal, *Les sous-groupes fermés de rang maximum des groupes de Lie clos.* Comment. Math. Helv. 23 (1949) 200–221. MR 11,326d

[14] A. Borovik and A. Nesin, *Groups of finite Morley rank.* Oxford University Press, New York (1994) xviii+409 pp. MR 96c:20004

[15] G. E. Bredon, *On homogeneous cohomology spheres.* Ann. of Math. 73 (1961) 556–565. MR 23#A243

[16] G. E. Bredon, *Introduction to compact transformation groups.* Academic Press, New York (1972) xiii+459 pp. MR 54#1265

[17] G. E. Bredon, *Sheaf theory. Second ed.* Springer GTM 170, Springer Verlag, NewYork (1997) xii+502 pp. MR 98g:55005, MR 36#4552

[18] K. Burns and R. Spatzier, *On topological Tits buildings and their classification.* Inst. Hautes Études Sci. Publ. Math. 65 (1987) 5–34. MR 88g:53049

[19] E. Cartan, *Sur des familles remarquables d'hypersurfaces isoparamétriques dans les espaces sphériques.* Math. Z. 45 (1939) 335–367. MR 1,28f

[20] H. Cartan and J.-P. Serre, *Espaces fibrés et groupes d'homotopie. II. Applications.* C. R. Acad. Sci. Paris 234 (1952) 393–395. MR 13,675b

[21] T. E. Cecil and P. J. Ryan, *Tight and taut immersions of manifolds.* Pitman Research Notes in Mathematics 107, Pitman, Boston, MA. (1985). vi+335 pp. MR 87b:53089

[22] S. Console and C. Olmos, *Clifford systems, algebraically constant second fundamental form and isoparametric hypersurfaces.* Manuscripta Math. 97 (1998) 335–342. MR 99j:53072

[23] T. tom Dieck, *Transformation groups.* de Gruyter Studies in Mathematics, 8, Berlin (1987) x+312 pp. MR 89c:57048

[24] A. Dold, *Lectures on algebraic topology.* Springer Verlag, New York-Berlin (1972) xi+377 pp. MR 54#3685

[25] J. Dorfmeister and E. Neher, *Isoparametric triple systems of FKM-type. I.* Abh. Math. Sem. Univ. Hamburg 53 (1983) 191–216. MR 85i:17002a

[26] J. Dorfmeister and E. Neher, *Isoparametric triple systems of FKM-type. II.* Manuscripta Math. 43 (1983) 13–44. MR 85i:17002b

[27] J. Dorfmeister and E. Neher, *Isoparametric triple systems of FKM-type. III.* Algebras Groups Geom. 1 (1984) 305–343. MR 86f:17005

[28] J. Dorfmeister and E. Neher, *Isoparametric hypersurfaces, case* $g = 6$, $m = 1$. Comm. Algebra 13 (1985) 2299–2368. MR 87d:53096

[29] E. B. Dynkin, *Semisimple subalgebras of semisimple Lie algebras.* Amer. Math. Soc. Transl. (2) 6 (1957) 111–244. MR 13,904c

[30] R. Engelking, *General topology.* Heldermann Verlag, Berlin, second edition (1989) viii+529 pp. MR 91c:54001

[31] J. H. Eschenburg and E. Heintze, *On the classification of polar representations.* Math. Z. 232 (1999) 391–398. MR 2001g:53099

[32] J.-H. Eschenburg and V. Schroeder, *Tits distance of Hadamard manifolds and isoparametric hypersurfaces.* Geom. Dedicata 40 (1991) 97–101. MR 92i:53038

[33] Y. Félix, S. Halperin and J.-C. Thomas, *Rational homotopy theory.* Springer GTM 205, Springer Verlag, New York (2001) xxv+535 pp.

[34] D. Ferus, H. Karcher, and H.F. Münzner, *Cliffordalgebren und neue isoparametrische Hyperflächen.* Math. Z. 177 (1981) 479–502. MR 83k:53075

[35] A. T. Fomenko, D. B. Fuchs, and V. L. Gutenmacher, *Homotopic topology.* Akadémiai Kiadó, Budapest (1986) 310 pp. MR 88f:55001

[36] K. Grove and S. Halperin, *Dupin hypersurfaces, group actions and the double mapping cylinder.* J. Differential Geom. 26 (1987) 429–459. MR 89h:53113

[37] Th. Grundhöfer and N. Knarr, *Topology in generalized quadrangles.* Topology Appl. 34 (1990) 139–152. MR 91c:51025

[38] Th. Grundhöfer, N. Knarr, and L. Kramer, *Flag-homogeneous compact connected polygons.* Geom. Dedicata 55 (1995) 95–114. MR 96a:51009

[39] Th. Grundhöfer, N. Knarr, and L. Kramer, *Flag-homogeneous compact connected polygons, II.* Geom. Dedicata 83 (2000) 1–29.

[40] Th. Grundhöfer, N. Knarr, and L. Kramer, *The classification of compact buildings.* Preprint Würzburg (1999).

[41] Th. Grundhöfer and R. Löwen, *Linear topological geometries.* In: *Handbook of incidence geometry,* p. 1255–1324. North-Holland, Amsterdam (1995). MR 97c:51008

[42] Th. Grundhöfer and H. Van Maldeghem, *Topological polygons and affine buildings of rank three.* Atti Sem. Mat. Fis. Univ. Modena 38 (1990) 459–479. MR 92c:51028

[43] E. Heintze and X. Liu, *Homogeneity of infinite-dimensional isoparametric submanifolds.* Ann. of Math. 149 (1999) 149–181. MR 2000c:58007

[44] K. H. Hofmann and S. A. Morris, *The structure of compact groups.* De Gruyter, Berlin (1998) xviii+835 pp. MR 99k:22001

[45] W.-y. Hsiang and H. B. Lawson, Jr., *Minimal submanifolds of low cohomogeneity.* J. Differential Geometry 5 (1971) 1–38. MR 45#7645

[46] W.-y. Hsiang, R. S. Palais, and C.-L. Terng, *The topology of isoparametric submanifolds.* J. Differential Geom. 27 (1988) 423–460. MR 89m:53104

[47] W.-y. Hsiang and J. C. Su, *On the classification of transitive effective actions on Stiefel manifolds.* Trans. Amer. Math. Soc. 130 (1968) 322-336. MR 36#4581

[48] S. Immervoll, *Smooth projective planes, smooth generalized quadrangles, and isoparametric hypersurfaces.* Ph.D. Thesis, Tübingen: Math. Fak., Univ. Tübingen, in preparation (2001).

[49] H. Karcher, *A geometric classification of positively curved symmetric spaces and the isoparametric construction of the Cayley plane.* Astérisque (163-164) 6 (1988) 111–135. MR 90g:53063

[50] B. Kleiner and B. Leeb, *Rigidity of quasi-isometries for symmetric spaces and Euclidean buildings.* Inst. Hautes Études Sci. Publ. Math. 86 (1997) 115–197. MR 98m:53068

[51] N. Knarr, *The nonexistence of certain topological polygons.* Forum Math. 2 (1990) 603–612. MR 92a:51018

[52] N. Knarr and L. Kramer, *Projective planes and isoparametric hypersurfaces.* Geom. Dedicata 58 (1995) 193–202. MR 96i:53059

[53] S. Kobayashi and K. Nomizu, *Foundations of differential geometry, Vol. I.* Interscience Publishers, New York-London (1963) xi+329 pp. MR 27#2945

[54] L. Kramer, *Compact polygons.* Ph.D. Thesis, Tübingen: Math. Fak., Univ. Tübingen, (1994) vi+72 pp. Zbl 844.51006
available as math.DG/0104064 at the Mathematics ArXiv,
`http://front.math.ucdavis.edu/math.DG/0104064`

[55] L. Kramer, *Octonion Hermitian quadrangles.* Bull. Belg. Math. Soc. Simon Stevin 5 (1998) 353–362. MR 99g:51003

[56] L. Kramer, *Compact homogeneous quadrangles and focal manifolds. Habilitationsschrift,* Würzburg: Math. Fak., Univ. Würzburg, (1998) 138 pp.

[57] L. Kramer, *Compact ovoids in quadrangles II. The classical quadrangles.* Geom. Dedicata 79 (2000) 179–188. MR 2001b:51007

[58] L. Kramer, *Compact ovoids in quadrangles III. Clifford algebras and isoparametric hypersurfaces.* Geom. Dedicata 79 (2000) 321–339.

[59] L. Kramer and H. Van Maldeghem, *Compact ovoids in quadrangles I. Geometric constructions.* Geom. Dedicata 78 (1999) 279–300. MR 2000m:51008

[60] L. Kramer, K. Tent, and H. Van Maldeghem, *Simple groups of finite Morley rank and Tits buildings.* Israel J. Math. 109 (1999) 189–224. MR 2000f:51022

[61] M. Kreck and S. Stolz, *A diffeomorphism classification of* 7*-dimensional homogeneous Einstein manifolds with* $SU(3) \times SU(2) \times U(1)$*-symmetry.* Ann. of Math. 127 (1988) 373–388. MR 89c:57042

[62] M. Kreck and S. Stolz, *Some nondiffeomorphic homeomorphic homogeneous* 7*-manifolds with positive sectional curvature.* J. Differential Geom. 33 (1991) 465–486. MR 92d:53043

[63] M. Kreck and S. Stolz, *A correction on: 'Some nondiffeomorphic homeomorphic homogeneous* 7*-manifolds with positive sectional curvature'.* J. Differential Geom. 49 (1998) 203–204. MR 99h:53069

[64] W. Kühnel and T. Cecil, *Bibliography on tight, taut and isoparametric submanifolds.* In: *Tight and taut submanifolds, Berkeley 1994*, p. 307–339, MSRI Publ. 32, Cambridge Univ. Press, Cambridge (1997).

[65] B. Leeb, *A characterization of irreducible symmetric spaces and Euclidean buildings of higher rank by their asymptotic geometry. Habilitationsschrift,* Bonn (1997) 42 pp.
Bonner Math. Schriften 326 (2000).

[66] R. Löwen, *Homogeneous compact projective planes.* J. Reine Angew. Mathematik 321 (1981) 217–220. MR 82c:51023

[67] L. N. Mann, *Dimensions of compact transformation groups.* Michigan Math. J. 14 (1967) 433–444. MR 36#3916

[68] E. Markert, *Isoparametric hypersurfaces and generalized quadrangles.* Diplomarbeit, Würzburg: Math. Inst., Univ. Würzburg (1999) 45 pp.

[69] J. McCleary, *User's guide to spectral sequences.* Publish or Perish Inc., Wilmington, DEL. (1985) xiv+423 pp. MR 87f:55014

[70] M. Mimura, *Homotopy theory of Lie groups.* In: *Handbook of algebraic topology*, p. 951–991. North-Holland, Amsterdam (1995). MR 97c:57038

[71] M. Mimura and H. Toda, *Topology of Lie groups. I, II.* Transl. of Math. Monographs 91, American Mathematical Society, Providence, RI. (1991) iv+451 pp. MR 92h:55001

[72] D. Montgomery, *Simply connected homogeneous spaces.* Proc. Amer. Math. Soc. 1 (1950) 467–469. MR 12,242c

[73] D. Montgomery and C. T. Yang, *Orbits of highest dimension.* Trans. Amer. Math. Soc. 87 (1958) 284–293. MR 20#6705

[74] D. Montogmery and L. Zippin, *Topological transformation groups.* Interscience Publishers, New York-London (1955) xi+282 pp. MR 17,282b

[75] G. D. Mostow, *Strong rigidity of locally symmetric spaces.* Princeton University Press, Princeton, NJ. (1973) v+195 pp. MR 52#5874

[76] H. F. Münzner, *Isoparametrische Hyperflächen in Sphären, I.* Math. Ann. 251 (1980) 57–71. MR 82a:53058

[77] H. F. Münzner, *Isoparametrische Hyperflächen in Sphären, II.* Math. Ann. 256 (1981) 215–232. MR 82m:53053

[78] F. Nietzsche, *Also sprach Zarathustra.* Deutscher Taschenbuch Verlag/de Gruyter (1988) 420 pp.

[79] C. Olmos, *Isoparametric submanifolds and their homogeneous structures.* J. Differential Geom. 38 (1993) 225–234. MR 95a:53102

[80] A. L. Onishchik, *Topology of transitive transformation groups.* Johann Ambrosius Barth Verlag GmbH, Leipzig (1994) xvi+300 pp. MR 95e:57058

[81] A. L. Onishchik and È. B. Vinberg, *Lie groups and algebraic groups.* Springer-Verlag, Berlin (1990). xx+328 pp. MR 91g:22001

[82] R. S. Palais and C.-L. Terng, *Critical point theory and submanifold geometry.* LNM 1353 Springer-Verlag, Berlin (1988). x+272 pp. MR 90c:53143

[83] J. Rohlfs and T. A. Springer, *Applications of buildings.* In: *Handbook of incidence geometry*, p. 1085–1114. North-Holland, Amsterdam (1995). MR 97b:20041

[84] H. Salzmann, *Homogene kompakte projektive Ebenen.* Pacific J. Math. 60 (1975) 217–234. MR 53#3878

[85] H. Salzmann, D. Betten, Th. Grundhöfer, H. Hähl, R. Löwen, and M. Stroppel, *Compact projective planes.* Walter de Gruyter & Co., Berlin (1995) xiv+688 pp. MR 97b:51009

[86] H. Scheerer, *Transitive actions on Hopf homogeneous spaces.* Manuscripta Math. 4 (1971) 99–134. MR 45#2086

[87] V. Schneider, *Transitive actions on highly connected spaces.* Proc. Amer. Math. Soc. 38 (1973) 179–185. MR 47#9658

[88] A. E. Schroth, *Topological circle planes and topological quadrangles.* Pitman Research Notes in Mathematics 337, Longman, Harlow (1995) x+155 pp. MR 97b:51010

[89] G. M. Seitz, *Maximal subgroups of exceptional algebraic groups.* Mem. Amer. Math. Soc. 90 (1991) iv+197 pp. MR 91g:20038

[90] E. H. Spanier, *Algebraic topology.* Springer-Verlag, New York-Berlin (1981). xvi+528 pp. MR 96a:55001

[91] S. Stolz, *Multiplicities of Dupin hypersurfaces.* Invent. Math. 138 (1999) 253–279. MR 2001d:53065

[92] Th. Strauß, *Cohomology rings, sphere bundles, and double mapping cylinders.* Diplomarbeit, Würzburg: Math. Inst., Univ. Würzburg, (1996) 45 pp.

[93] M. Stroppel, *Reconstruction of incidence geometries from groups of automorphisms.* Arch. Math. 58 (1992) 621–624. MR 93e:51026

[94] M. Stroppel, *A categorical glimpse at the reconstruction of geometries.* Geom. Dedicata 46 (1993) 47–60. MR 94c:51036

[95] W. Strübing, *Isoparametric submanifolds.* Geom. Dedicata 20 (1986) 367–387. MR 87j:53084

[96] J. Szenthe, *On the topological characterization of transitive Lie group actions.* Acta Sci. Math. (Szeged) 36 (1974) 323–344. MR 50#13368

[97] R. Takagi, *A class of hypersurfaces with constant principal curvatures in a sphere.* J. Differential Geom. 11 (1976) 225–233. MR 54#13798

[98] R. Takagi and T. Takahashi, *On the principal curvatures of homogeneous hypersurfaces in a sphere.* In: *Differential geometry in honor of K. Yano*, p. 469–481, Kinokuniya, Tokyo (1972). MR 48#12413

[99] K. Tent, *Very homogeneous generalized polygons of finite Morley rank.* J. London Math. Soc. 62 (2000) 1–15. MR 2001f:51011

[100] C.-L. Terng, *Isoparametric submanifolds and their Coxeter groups.* J. Differential Geom. 21 (1985) 79–107. MR 87e:53095

[101] G. Thorbergsson, *Isoparametric foliations and their buildings.* Ann. of Math. 133 (1991) 429–446. MR 92d:53053

[102] G. Thorbergsson, *Clifford algebras and polar planes.* Duke Math. J. 67 (1992) 627–632. MR 93i:51033

[103] G. Thorbergsson, *A survey on isoparametric hypersurfaces and their generalizations.* In: *Handbook of differential geometry, I*, P. 963–995 North Holland, Amsterdam (2000). MR 2001a:53097

[104] J. Tits, *Tabellen zu den einfachen Lie Gruppen und ihren Darstellungen.* LNM 40, Springer-Verlag, Berlin (1967). v+53 pp. MR 36#1575

[105] J. Tits, *Représentations linéaires irréductibles d'un groupe réductif sur un corps quelconque.* J. Reine Angew. Math. 247 (1971) 196–220. MR 43#3269

[106] J. Tits, *Buildings of spherical type and finite BN-pairs.* LNM 386, Springer-Verlag, Berlin (1974). x+299 pp. MR 57#9866

[107] J. Tits, *Endliche Spiegelungsgruppen, die als Weylgruppen auftreten.* Invent. Math. 43 (1977) 283–295. MR 57#478

[108] J. Tits and R. Weiss, *Moufang polygons.* Book in preparation (2001).

[109] F. Uchida, *An orthogonal transformation group of* $(8k-1)$*-sphere.* J. Differential Geom. 15 (1980) 569–574. MR 83a:57056

[110] H. Van Maldeghem, *Generalized polygons.* Birkhäuser Verlag, Basel (1998). xvi+502 pp. MR 2000k:5104

[111] H. C. Wang, *Homogeneous spaces with non-vanishing Euler characteristics.* Ann. Math. 50 (1949) 925–953. MR 11,326c

[112] Q. M. Wang, *On the topology of Clifford isoparametric hypersurfaces.* J. Differential Geom. 27 (1988) 55–66. MR 89e:53093

[113] G. W. Whitehead, *Elements of homotopy theory.* Springer-Verlag, New York-Berlin (1978). xxi+744 pp. MR 80b:55001

[114] M. Wolfrom, Ph.D. Thesis, Würzburg: Math. Fak., Univ. Würzburg, in preparation (2001).

Editorial Information

To be published in the *Memoirs*, a paper must be correct, new, nontrivial, and significant. Further, it must be well written and of interest to a substantial number of mathematicians. Piecemeal results, such as an inconclusive step toward an unproved major theorem or a minor variation on a known result, are in general not acceptable for publication. Papers appearing in *Memoirs* are generally longer than those appearing in *Transactions*, which shares the same editorial committee.

As of February 28, 2002, the backlog for this journal was approximately 4 volumes. This estimate is the result of dividing the number of manuscripts for this journal in the Providence office that have not yet gone to the printer on the above date by the average number of monographs per volume over the previous twelve months, reduced by the number of volumes published in four months (the time necessary for preparing a volume for the printer). (There are 6 volumes per year, each containing at least 4 numbers.)

A Consent to Publish and Copyright Agreement is required before a paper will be published in the *Memoirs*. After a paper is accepted for publication, the Providence office will send a Consent to Publish and Copyright Agreement to all authors of the paper. By submitting a paper to the *Memoirs*, authors certify that the results have not been submitted to nor are they under consideration for publication by another journal, conference proceedings, or similar publication.

Information for Authors

Memoirs are printed from camera copy fully prepared by the author. This means that the finished book will look exactly like the copy submitted.

The paper must contain a *descriptive title* and an *abstract* that summarizes the article in language suitable for workers in the general field (algebra, analysis, etc.). The *descriptive title* should be short, but informative; useless or vague phrases such as "some remarks about" or "concerning" should be avoided. The *abstract* should be at least one complete sentence, and at most 300 words. Included with the footnotes to the paper should be the 2000 *Mathematics Subject Classification* representing the primary and secondary subjects of the article. The classifications are accessible from `www.ams.org/msc/`. The list of classifications is also available in print starting with the 1999 annual index of *Mathematical Reviews*. The Mathematics Subject Classification footnote may be followed by a list of *key words and phrases* describing the subject matter of the article and taken from it. Journal abbreviations used in bibliographies are listed in the latest *Mathematical Reviews* annual index. The series abbreviations are also accessible from `www.ams.org/publications/`. To help in preparing and verifying references, the AMS offers MR Lookup, a Reference Tool for Linking, at `www.ams.org/mrlookup/`. When the manuscript is submitted, authors should supply the editor with electronic addresses if available. These will be printed after the postal address at the end of the article.

Electronically prepared manuscripts. The AMS encourages electronically prepared manuscripts, with a strong preference for $\mathcal{A}_{\mathcal{M}}\mathcal{S}$-LaTeX. To this end, the Society has prepared $\mathcal{A}_{\mathcal{M}}\mathcal{S}$-LaTeX author packages for each AMS publication. Author packages include instructions for preparing electronic manuscripts, the *AMS Author Handbook*, samples, and a style file that generates the particular design specifications of that publication series. Though $\mathcal{A}_{\mathcal{M}}\mathcal{S}$-LaTeX is the highly preferred format of TeX, author packages are also available in $\mathcal{A}_{\mathcal{M}}\mathcal{S}$-TeX.

Authors may retrieve an author package from e-MATH starting from `www.ams.org/tex/` or via FTP to `ftp.ams.org` (login as `anonymous`, enter username as password, and type `cd pub/author-info`). The *AMS Author Handbook* and the *Instruction Manual* are available in PDF format following the author packages link from `www.ams.org/tex/`. The author package can be obtained free of charge by sending email to `pub@ams.org` (Internet) or from the Publication Division, American Mathematical Society, P.O. Box 6248, Providence, RI 02940-6248. When requesting an author package, please specify $\mathcal{A}_{\mathcal{M}}\mathcal{S}$-LaTeX or $\mathcal{A}_{\mathcal{M}}\mathcal{S}$-TeX, Macintosh or IBM (3.5) format, and the publication in which your paper will appear. Please be sure to include your complete mailing address.

Sending electronic files. After acceptance, the source file(s) should be sent to the Providence office (this includes any TeX source file, any graphics files, and the DVI or PostScript file).

Before sending the source file, be sure you have proofread your paper carefully. The files you send must be the EXACT files used to generate the proof copy that was accepted for publication. For all publications, authors are required to send a printed copy of their paper, which exactly matches the copy approved for publication, along with any graphics that will appear in the paper.

TeX files may be submitted by email, FTP, or on diskette. The DVI file(s) and PostScript files should be submitted only by FTP or on diskette unless they are encoded properly to submit through email. (DVI files are binary and PostScript files tend to be very large.)

Electronically prepared manuscripts can be sent via email to `pub-submit@ams.org` (Internet). The subject line of the message should include the publication code to identify it as a Memoir. TeX source files, DVI files, and PostScript files can be transferred over the Internet by FTP to the Internet node `e-math.ams.org` (130.44.1.100).

Electronic graphics. Comprehensive instructions on preparing graphics are available at `www.ams.org/jourhtml/graphics.html`. A few of the major requirements are given here.

Submit files for graphics as EPS (Encapsulated PostScript) files. This includes graphics originated via a graphics application as well as scanned photographs or other computer-generated images. If this is not possible, TIFF files are acceptable as long as they can be opened in Adobe Photoshop or Illustrator. No matter what method was used to produce the graphic, it is necessary to provide a paper copy to the AMS.

Authors using graphics packages for the creation of electronic art should also avoid the use of any lines thinner than 0.5 points in width. Many graphics packages allow the user to specify a "hairline" for a very thin line. Hairlines often look acceptable when proofed on a typical laser printer. However, when produced on a high-resolution laser imagesetter, hairlines become nearly invisible and will be lost entirely in the final printing process.

Screens should be set to values between 15% and 85%. Screens which fall outside of this range are too light or too dark to print correctly. Variations of screens within a graphic should be no less than 10%.

Inquiries. Any inquiries concerning a paper that has been accepted for publication should be sent directly to the Electronic Prepress Department, American Mathematical Society, P. O. Box 6248, Providence, RI 02940-6248.

Editors

Selected Titles in This Series

(Continued from the front of this publication)